Advances in Surface Modification and Treatment of Wood

Advances in Surface Modification and Treatment of Wood

Editor

Levente Csóka

MDPI • Basel • Beijing • Wuhan • Barcelona • Belgrade • Manchester • Tokyo • Cluj • Tianjin

Editor
Levente Csóka
Celltech-paper Ltd.
Hungary

Editorial Office
MDPI
St. Alban-Anlage 66
4052 Basel, Switzerland

This is a reprint of articles from the Special Issue published online in the open access journal *Coatings* (ISSN 2079-6412) (available at: https://www.mdpi.com/journal/coatings/special_issues/surface_wood).

For citation purposes, cite each article independently as indicated on the article page online and as indicated below:

LastName, A.A.; LastName, B.B.; LastName, C.C. Article Title. *Journal Name* **Year**, *Volume Number*, Page Range.

ISBN 978-3-0365-0902-0 (Hbk)
ISBN 978-3-0365-0903-7 (PDF)

Cover image courtesy of Carlos A. García-Gonzálezc.

Contents

About the Editor

Levente Csóka (Professor, 45 years old, Doctor of the Hungarian Academy of Science, he holds the Hungarian Order of Merit (Knight's Cross)) has extended research experience in the field of cellulose, pulp and paper manufacturing, nanoparticle processing and testing, wastewater remediation and ultrasound extraction of bioactive compounds from hemp—both for fundamental and industrial R&D projects. He is one of the pioneers of using nanotechnology in the paper industry, including layer-by-layer coatings and manufacturing of antimicrobial nanoparticles. He has held positions in Japan, the USA and Finland. He is co-author of more than 200 publications and has an H index of 20.

Preface to "Advances in Surface Modification and Treatment of Wood"

Timber is a biosynthetic end-product. The photosynthetic formation of wood is a function of the gene expression and catalytic rates of structural enzymes under different environmental factors. Thus, to achieve a full understanding of wood formation, as a later subject of our studies, each component of the full set of intrinsic processes must be studied (i.e., chemical reactions and physical changes) and known, investigating how each one of those components is affected by other processes in a complex form. Differences in the wood surfaces occur between the manufacturing and post-treatment processes within single trees. Wood surface attributes can be established by examining its several different physical or chemical properties. Understanding how their unique anisotropic molecular organization, chemical linkages, branching, and other molecular features govern micro- and macroscale accessibility is essential for coating and complex modification processes. It is therefore important for scientific as well as practical reasons to qualify and quantify the effects of wood surface treatments and modifications.This Special Issue contains nine scientific works from a wide spectrum of research with a specific focus on the surface modifications, potential applications, and natural-artificial weathering treatments. The interfacial interactions of the wood surface components are key concerns for the reliability of structural changes after modifications. Wood surface modification is a comprehensive concept, which, in time, has been as successful as challenging in terms of improving the resistance during its life cycle in both indoor and outdoor applications. The review paper in this Special Issue offers an systematic overview of some of the most recent advances reported in the field. Another study deals with the effects of nanosilver impregnation and heat treatment on the gas permeability and pull-off strength, as these properties depend to some extent on the porous structure of the wood species. Impregnation can increase the thermal conductivity of the specimens and decrease the heat treatment gradient. The structural and fundamental changes occurring in naturally aged wood surfaces was the objective of another study to compare the performance of transparent and pigmented coating systems. To follow artificial ageing in different wood species' using surface chemical compounds without coatings, a novel evaluation method was developed. Confocal profilometry and macro-photography were used to quantify the image shelling of flat-faced deckboards exposed to natural weathering for seven months. A different study was dedicated to examining the changes in the adhesion strength between transparent acrylic and alkyd coatings on wood due to weathering changes. In parallel, some other impacts were evaluated, including the presence of hindered amine light stabilizer or hydroxyphenyl-benzotriazoles UV additives in coatings and the fungicidal pre-treatment of wood with boric acid or benzalkonium chloride, as well as plasma modification. A following study examined how the overall service life of the acrylate and oil coating system is affected by the initial surface modification of wood using caffeine solutions, dispersion of TiO_2 nanoparticles, and their combinations. The method of artificial-accelerated weathering in a UV chamber was used with a subsequent test of mold growth on both ageless and weathered surfaces of treated wood. The authors observed that, in service, wood treated with copper-based preservatives tends to change from green/blue to a brown color as the surface of the wood oxidizes. The aim of the following scientific contribution was to understand the efficacy of this approach in yielding a brown, photo-stable wood surface with the potential to eliminate or reduce the need for the addition of colorants. Finally, spruce with different heat pre-treatments was studied with the effect of diffuse coplanar surface barrier discharge plasma treatment, which was evaluated by measuring the changes in wood surface free energy, chemical composition, and micro-morphology.Challenges are still exist to fully understanding the effect of the numerous applied chemicals and the wide range of treatment processes on wood surfaces, which can be overcome to some extent by further collaborative efforts between scientist, laboratories, and industries.

Levente Csóka
Editor

coatings

Review

Wood Surface Modification—Classic and Modern Approaches in Wood Chemical Treatment by Esterification Reactions

Carmen-Alice Teacă [1,*] and Fulga Tanasă [2]

1 Centre of Advanced Research in Bionanoconjugates and Biopolymers, "Petru Poni" Institute of Macromolecular Chemistry, 41A Grigore Ghica-Vodă Alley, 700487 Iaşi, Romania
2 Polyaddition and Photochemistry Department, "Petru Poni" Institute of Macromolecular Chemistry, 41A Grigore Ghica-Vodă Alley, 700487 Iaşi, Romania; ftanasa@icmpp.ro
* Correspondence: cateaca@icmpp.ro or cateaca14@yahoo.com

Received: 15 June 2020; Accepted: 29 June 2020; Published: 30 June 2020

Abstract: Wood surface modification is a comprehensive concept which, in time, turned out to be as successful as challenging when it comes to improve the resistance of wood during its life cycle in both indoor and outdoor applications. The initial approaches have aimed at simple methods with immediate results. Nowadays, the paradigm has slightly changed due to the scientific and technical advances, and some methods has become intermediate stages in more complex processes, after being used, for long time, as stand-alone procedures. The esterification was employed as a convenient method for wood surface modification due to the high amount of free hydroxyl groups available at the surface of wood and other lignocellulosic materials. Therefore, different esterification approaches were tested: activated condensation with carboxylic acids (monocarboxylic, as well as dicarboxylic acids, fatty acids, etc.) in the presence of condensation activating agents (such as trifluoroacetic anhydride); reaction with β-halogen-substituted carboxylic acids; esterification using carboxylic acids derivatives (acyl chlorides, anhydrides) or even multifunctional carboxylic acids (i.e., tricine). Thus, wood with improved dimensional stability and weathering resilience, higher fire resistance, enhanced hydrophobic character, and mechanical durability was obtained. This paper offers an overview of some of the most recent advances reported in the field, presented in a systematic manner, using the type of reaction as classification criterion. The main improvements will be outlined in a critical assessment in order to provide an useful tool for a wise choice in future applications.

Keywords: wood; surface modification; esterification; classic approaches; modern approaches

---

1. Introduction

Wood surface modification is a comprehensive concept which, in time, turned out to be as increasingly successful as challenging when it comes to turn wood species into high-value products. Thus, further improvement in the wood resistance during its life cycle in both indoor and outdoor applications can be attained.

Appropriately modified products of wood and wood-based composites are increasingly being applied in housing, the building industry, transport etc.

Many approaches envisaging wood modification [1–8] include treatment processes, defined as chemical, biological (e.g., enzymatic modification for hydrophobicity), mechanical or physical methods. All these methods are applied to modify wood with the scope to acquire enhanced properties for the sustainable service life of the resulted material which should not pose any environmental hazards related to its toxicity during all life cycle period (referring to use, recycling or disposal issues). Other modification methods are suitable for increasing wood properties such as dimensional stability and durability (resistance to biological degradation, UV radiation, humidity, and harmful chemicals; thermal stability or fire resistance; mechanical properties), among these impregnation and coating treatment approaches being often applied.

Impregnation treatment represents a passive modification strategy which does not change the molecular structure of wood polymer constituents. It involves usually diffusion of bulking chemicals, such as monomers, polymers, resins and waxes, into either cell wall or lumen, and determines increasing wood density and stop water penetration in the cell walls.

Coating modification strategy implies formation of a physical barrier layer consisting of moisture-, bio-, fire- or UV-resistant agents on the wood surfaces, as the final stage of wood processing. It is effective for an improved wood protection against weathering processes under environmental exposure conditions in exterior uses. Nevertheless, implementation of new alternatives through exploiting new agents derived from natural resources (e.g., resins, waxes, biopolymers) for wood treatment may be of real significance when issues such as reducing the cost and environmental risk at the end-of-life of treated wood are considered [9]. The main strategies applied for wood treatment are schematically represented in Figure 1.

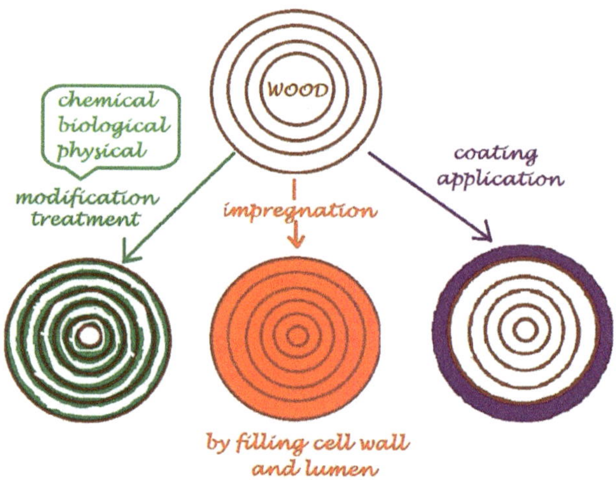

Figure 1. Schematic representation of the methods used for wood modification.

The wood tree chemistry approach [10] may be valuable for investigation of the chemical modification performed both on the wood [2,5] and its major polymer components (cellulose and lignin). Through this process, materials with given and special properties (e.g., thermal stability [11]), resistance to decay under biological organisms' action [12–17] and UV exposure [18–24] may be obtained. Beside all the above-mentioned, the introduction of some reactive functional groups through addition of different chemicals to the hydroxyl groups from wood material can be successfully applied for further grafting synthetic polymers in order to obtain composites [25–29].

Wood fibers surfaces are chemically reactive due to the presence of many hydroxyl groups in the structure of their main polymer constituents, namely lignin and cellulose, thus becoming susceptible for efficient surface modification treatments [30,31]. The inherent reactivity of these hydroxyl groups can be harnessed through modification methods which improve their interfacial interactions with non-polar polymer matrices in composite formulations. In such obtained multi-component polymer systems are consequently noticed improved properties, including wettability, gluing and adhesion of resin type coatings in the case of wood acetylation for example [32], through modifying the surface energy and polarity of wood surfaces [33]. Wood modification strategies include different approaches [34] such as:

- Physical methods: electrical discharge-corona, cold plasma treatments [35–39], thermal treatment [40] and mercerization [41];
- Chemical methods: pre-treatment of wood fibers surface, grafting, use of coupling agents and functionalized polyolefin-coupling agents [42].

Chemical modification of wood is based on the reaction of the wood structural polymer components with chemical reactants that do not have any protective result (e.g., biocides or fire retardants) but which determines the required protective effect and improvement in selected properties directly in the wood substrate. The chemical reactants used for wood modification can act on the surface on which the remain or only partly penetrate wood superficial layers (namely, passive modification strategies such as impregnation), but they also can be present deeper in the wood structure (in the lumens of the cells or inside the cell walls) when they react through chemical bonding with the polysaccharides and lignin from wood (namely, active modification strategies).The new interactions that occur between the wood surfaces and modifying agents are of chemical or only mechanical nature, these contributing to the increased resistance of the modified wood against water, UV radiation and biological pests. Thermosetting polymers (e.g., resins such as phenolic, amino and furfuryl-alcohol type) and thermoplastic polyacrylates are usually preferred for filling the lumens of wood cells during wood modification. Occurrence of reactions with the hydroxyl groups or other chemical groups of the wood when using reactants (e.g., carboxylic acids and their anhydrides, aldehydes, lactones, isocyanates, nitriles, or epoxides) can generate formation of reactive monomers [2,3].Some usual methods often applied for modification of wood surfaces, through both active and passive ways, can be summarized as presented in Table 1.

Table 1. Examples of methods used for wood modification.

Type of Treatment	Reference
Isocyanate treatment (ease reaction in the presence of pyridine or acidic catalysts)	[43,44]
Acrylation	[45]
Benzoylation	[23,46,47]
Acid chlorides treatment (ex. octanoyl chloride and palmitoyl chloride)	[28]
Acid anhydride treatment: -acetic anhydride -maleic anhydride -succinic anhydride	[2,4] [48,49] [21,25,50–52]
Silane treatment	[16,53]
Furfurylation treatment	[54–59]
Ketene treatment (usually applied for wood acetylation in order to avoid formation of acetic acid as by-product)	[60–63]
Other anhydrides: ex. crotonic, propionic or methacrylic anhydride	[64,65]
Polycarboxylic acids treatment:1,2,3,4-butanetetracarboxylic acid, citric acid (through impregnation)	[29,66–73]
Tricine treatment (tricine is a zwitterionic aminoacid)	[74,75]
1,3-Dimethylol-4,5-dihydroxyethyleneurea (DMDHEU) treatment	[76–80]
Fatty acids treatment	[81]
Fatty acid chlorides treatment (induce thermo-plasticity into wood)	[82,83]
Oxalic acid and cetyl alcohol treatment	[84]
Isopropenyl acetate (in the presence of anhydrous aluminum chloride as a catalyst)	[13]

Most of these modification strategies are in fact esterification approaches and some of them will be discussed below.

2. Wood Surface Modification by Esterification Reactions Using Anhydrides

The reaction of wood subjected to anhydride treatment (e.g., acetic anhydride, maleic anhydride, succinic anhydride) can be exemplified as presented in Scheme 1. Through such chemical treatment, the molecular structure of wood polymer constituents is changed with positive effects on the wood properties, mainly the hydrophilic characteristics, thus the wood surface becomes more hydrophobic. Chemical modification of wood fibers is performed to provide a good compatibility by reaction with anhydrides that blocks the hydroxyl groups in the wood chemical structure, mainly for further inclusion in composite formulations with different polymer matrices. All these changes result in an improvement of biological resistance against various pests, and consequently have favorable impact as enhancing the dimensional stability and strength properties [85–87]. Cyclic anhydrides may react partially with wood when results a single ester function and a free carboxylic group.

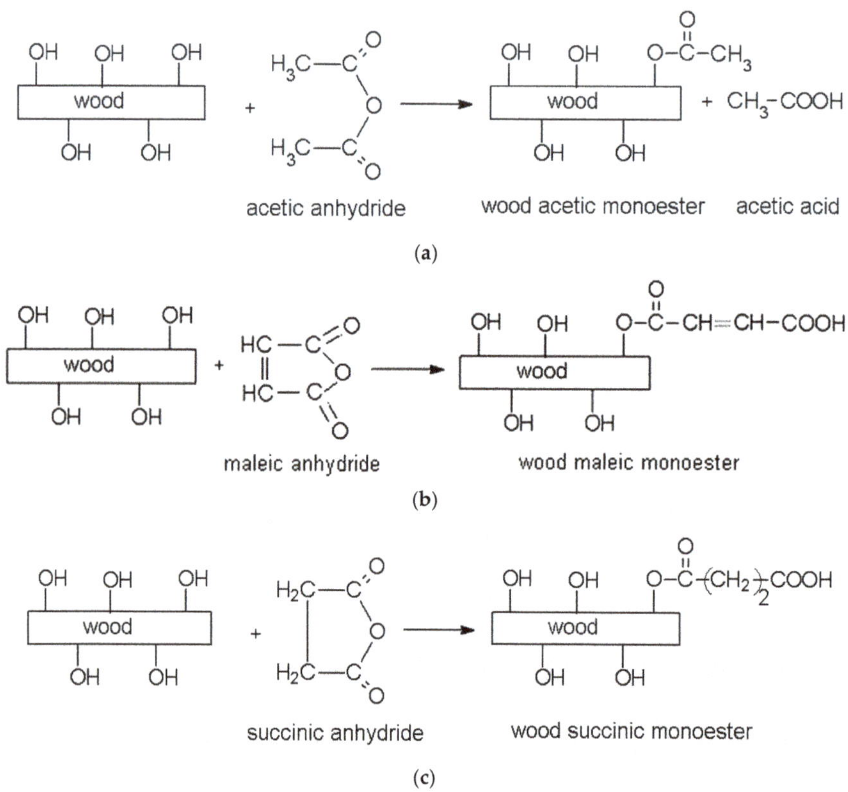

Scheme 1. Reaction of wood with organic anhydrides: linear (**a**) and cyclic—(**b**) and (**c**).

Acetylation treatment of wood involves the substitution of hydrophilic hydroxyl groups with hydrophobic acetyl groups [2]. At the same time, an increase in the dimensions of the acetylated wood substrates is observed because of swelling of the wood cell wall. In most cases, chemically modified wood presents a lower affinity for water absorption, comparatively with the non-modified wood. By replacing some of the hydroxyl groups on the wood polymers with acid anhydride, the hygroscopic properties of the wood are reduced with positive effects on the resulted properties when combined with other polymer matrices in composite formulations.

Usually, the chemical modification confers wood dimensional stability by deposition of the modifying reactant in the wood cell wall (bulking effect), and/or by cross-linking the wood cell wall polymers [88,89]. The beneficial effects of using acetylation for wood surface modification include the improvement in the weathering and coating performance [90].

When cyclic anhydrides are used for wood chemical treatment, e.g., succinic anhydride, or maleic anhydride, the carboxylic acid formed through esterification reaction is attached to the wood, as presented in Scheme 1b,c. Nevertheless, an extra cross-linking process can stabilize wood materials even better (e.g., improvement in dimensional stability and increased resistance to biological pests), besides improving the hydrophobic properties [3].

Our previous studies regarding the wood substrates and wood constituent polymers modification by esterification reaction using carboxylic acid anhydrides aimed to investigate the effect of such treatments on the wood thermal stability [11,51], on the structural changes occurred in modified wood [49–51] and on its photo-stability behavior under exposure to the artificial light irradiation [21], as well as on the structural and properties changes induced when modified wood is included in composite formulations comprising thermoplastic polymer matrix such as polyvinyl chloride (PVC) [27].

Fourier transform infrared spectroscopy (FT-IR) is a useful technique for studying the chemical and structural changes that occur in wood and wood polymer components due to esterification treatments [49]. The extent of reaction can be calculated as weight percent gain (WPG) determined by the differences in dry weight of the wood substrate before modification (W_1) and after modification (W_2) according to the equation [WPG = $(W_2 - W_1)/W_1 \times 100\%$]. A reaction parameter, namely time is very significant when consider evaluation of extent of reaction by calculating WPG values, but also it is important the method used for esterification (solvent, anhydride concentration). The amount of water present in the wood substrates is also essential, because a moisture level of around 5% appears to be necessary for an esterification reaction in optimal conditions, while above this value the present water hydrolyses the anhydrides with formation of the corresponding carboxylic acid. It seems that for each 1% of water content in wood sample, a loss of 5.7% of the modifying anhydride can be observed. An inverse correlation exists between the rate of esterification reaction and the moisture content in wood substrates.

Differences in chemical alterations that occurred in wood substrates and their main polymer components (cellulose, lignin) after modification using anhydrides are usually evidenced through analyzing the FTIR spectra, as exemplified in Figure 2, where one can observe the structural units that undergo various changes. These are functional groups located on the glucose monomer in cellulose chains from wood structure, when the carbon atoms occupying various positions in the glucopyranose ring (denoted as C-1, C-2, … C-6) are losing their identity, gradually being transformed into various carbonyl groups of different degrees of freedom, namely ketonic, aldehydic, and carboxylic groups [49].

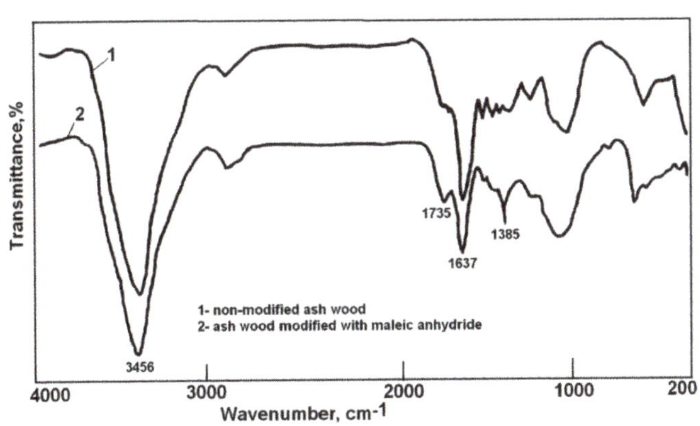

Figure 2. FTIR spectra recorded for wood sample from hardwood tree species (1—non-modified; 2—modified with maleic anhydride)—re-drawn from [49].

As previously presented in one of our first papers dealing with wood modification using anhydrides [49], the decrease in the absorption band for OH groups observed at 3456 cm^{-1} is an indicator of the reduction in hydroxyl group content in the hardwood sample after reaction with anhydride. A strong carbonyl band is noticed at 1735 cm^{-1} (attributed to the formation of C=O ester bonds), this corresponding mainly to the higher xylan content (a hemicellulose-type component) in hardwood sample, as well as an increase in the intensity of OH in plane bending vibration at 1385 cm^{-1} band which is specific to the wood polysaccharide components, namely cellulose and hemicelluloses.

Weak absorption bands between 1500 and 1400 cm^{-1} are specific to the aromatic ring vibrations and ring breathing with C–O groups stretching in lignin polymer from wood substrate. The evidence that modified wood is free both of un-reacted anhydride and of the by-product of carboxylic acid is given by the absence of absorption in the 1800–1760 cm^{-1} range, and lack of absorption band at 1700 cm^{-1}, respectively.

Usually, the wood substrates subjected to chemical modification through esterification using anhydrides have to be free of extractives in order to reduce the influence of such compounds on the reaction course and for a better yield. Extractives-free wood is also the supposition for the determination of wood structural constituents like holocellulose (which include all polysaccharides from wood), cellulose, hemicelluloses and lignin accordingly to the Technical Association of the Pulp and Paper Industry (TAPPI) standard methods. The cold water, hot water and organic solvents (toluene-ethanol mixture; xylene) extractions can all greatly improve the permeability of wood cell walls for its further treatments, including esterification with anhydrides [11,25,27,48–51]. Wood fibers esterified, for example, with maleic anhydride and used to prepare composites present reduced water absorption of both of fibers and the resulting composites in comparison to the systems comprising non-modified wood fibers [91–94]. At the same time, properties such as thermal behavior can be affected by the addition of the modified wood and wood constituent polymers resulted by reaction with maleic anhydride in composites with thermoplastic like polyvinyl chloride (PVC) [27]. Modified wood substrates are more prone to degradation with increasing anhydride concentration value. Weight loss values for the main decomposition temperature domain decrease with increasing the modified component (wood, wood polymers) amount in composites. Thermal behavior is improved mainly for the composite samples with modified lignin in composition. As expected, the composites with modified cellulose present a lower thermal stability by comparison with those comprising modified wood and modified lignin, respectively.

3. Wood Surface Modification by Tosylation Reaction

Wood modification, whatever the method, is aiming at altering (especially improving) the properties of the material, either at the surface or in depth. A better level of performance—dimensional stability, mechanical strength, weathering resistance, resilience under biological attack—can be achieved by changing the wood structure at the cell wall level.

The chemical modification remains one of the most used methods as it enables long term improvements through the formation of strong, stable covalent bonds between wood constituents (cellulose, lignin, hemicelluloses) and chemical reagents [95–98], despite its inherent drawbacks (such as: high production costs, use of toxic reagents and solvents, necessity to dispose in a controlled manner of the toxic residues, etc.). The term "wood chemical modification" was used for the first time in 1945 [99], and since then many reagents have been tested for this purpose: anhydrides-acetic, butyric, phthalic, succinic, maleic, propionic; acid chlorides; alkyl chlorides; ketene; β-propiolactone; mono-/dicarboxylic acids; different isocyanates; aldehydes-formaldehyde, acetaldehyde; difunctional aldehydes-trichloroacetaldehyde; o-phthalaldehydic acid; acrylonitrile; dimethyl sulfate; epoxides-ethylene, propylene, and butylene oxide and difunctional epoxides [4,100].

Wood acetylation with acetic anhydride is a classic esterification method that takes place at the available hydroxilic groups in wood. It was carried out, at the beginning, in the presence of zinc chloride ($ZnCl_2$) and pyridine (Py) as catalytic system [101], and later in liquid phase (homogeneous system) [96,102], or even without catalysts in more recent approaches [103,104]. Acetylated wood has a high dimensional stability and enhanced decay resistance, despite the residual acetic acid retained that can cause the corrosion of the ferrous elements [105]. Following the same idea, studies have focused also on the use of other esterification agents, such as methane sulfonic acid and p-toluene sulfonic acid, when the resulted esterified materials may be used as such or submitted to further modification reactions.

In a classic approach, the tosylation was carried out in liquid phase (N,N-dimethylacetamide, DMAC, and lithium chloride, LiCl), in the presence of triethylamine (TEA) as acid chloride scavenger and to prevent the occurrence of undesirable side reactions, for 24 h, at 8 °C, where tosyl chloride (TsC1) was employed as reagent [2]. Basically, the wood dried samples are immersed in the reaction medium containing all reagents and solvents. As result, available –OH groups of the wood cell walls were esterified with p-toluene sulfonyl moieties and, by the means of the nuclear magnetic resonance spectroscopy (NMR) study, it was possible to estimate that esterification at the oxygen atom at the C-6 carbon atom evolved with the highest rate as compared to those at C-2 and C-3 of the glucose ring (see Figure 3).

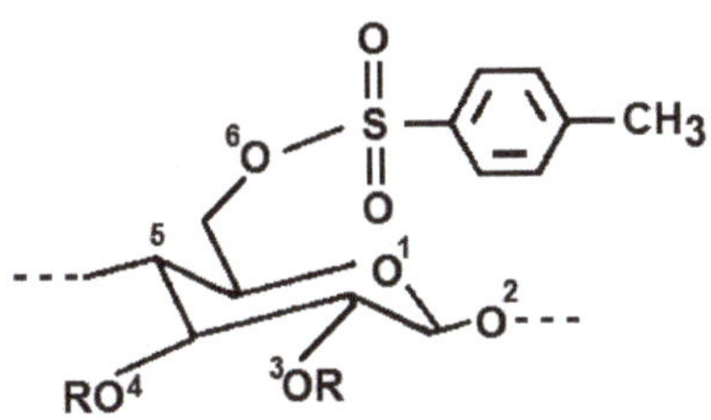

Figure 3. The structure of cellulose tosylate where the ester bond is at the O-6 atom.

Also, it was proven that the treatment entailed the reduced hydrophilic character of wood and an enhanced dimensional stability as effect of the presence of a bulky substituent in the cell walls [106].

If the sequence of operations is modified, the procedure implies the use of previously dried wood samples which are immersed in pyridine and allowed to swell for 24 h; then, the tosyl chloride solution is added along with other reagents and allowed to react for 24 h at 5 °C [107].

The reaction evolves by bonding each unit of TsCl with one available –OH group, so the amount of TsCl necessary is calculated taking into consideration the anhydrous glucopyranose equivalent (AGU). This procedure allows a higher weight increase, which means a higher amount of tosyl units (15%) were linked to the wood cell walls. Further investigations showed that tosylated wood has an enhanced permeability toward hydrophobic agents under mild conditions which makes it fit for subsequent processing [106].

When pure cellulose was submitted to such chemical modification, the corresponding tosylates are readily soluble in some of the most commonly used organic solvents (dimethyl sulfoxide—DMSO, N,N-dimethylacetamide—DMAC, N,N-dimethylformamide—DMF, acetone, tetrahydrofuran—THF, chloroform, etc.), depending on their degree of substitution [2]. In effect, numerous studies have been focusing on testing various solvents and mixtures able to allow high esterification yields, with or without derivatization, using activation as intermediate stage, either in homogeneous or heterogeneous systems [108].

Apart from –OH groups in cellulose, other available hydroxyl moieties are those from lignin and hemicelluloses. In the case of lignin tosylation, the mechanism has some distinct characteristics due to its aromatic structure and cross-linked supramolecular architecture [109]. Although ether bridges β-O-4 and α-O-4 are susceptible to be thermally and/or chemically broken, it is possible still to perform the tosylation of the primary hydroxyl groups (Scheme 2), especially from guaiacyl and syringyl sequences in lignin, under weak alkaline/neutral conditions, and in time intervals ranging between 5–6 up to 24 h [110–112].

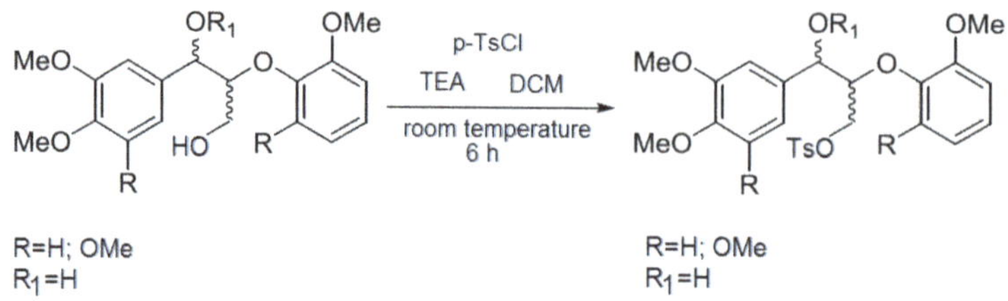

Scheme 2. Tosylation of primary –OH group at a β-O-4 bridge (benzylic position) in lignin (TEA = triethylamine; DCM = dichloromethane).

In a typical case study, lignin was reacted with p-toluenesulfonyl chloride in aqueous medium, in the presence of TEA, at 25°, when the tosylation reaction was completed in 24 h [111], without using other solvents or activating agents such as N,N-dicyclohexylcarbodiimide (DCC) and 4-pyrrolidinonepyridine (PP) [113], 4-dimethylaminopyridine (DMAP) or N,N-diisopropylcarbodiimide (DIC) [114,115]. The degree of substitution can be modulated by setting the TsCl/OH lignin ratio, depending on the final use of the tosylated material. Even more, at higher TsCl content, the tosylated lignin is less thermally stable as there are only a few free –OH groups available for thermal condensation of lignin fragments [116].

Modern approaches have performed the tosylation in the presence of ionic liquids (ILs). In the past decades, they have been used as highly efficient green solvents for ligno-cellulosic materials: fractionation and biorefinery, dissolution of polysaccharides, processing cellulosic fibers, and, particularly, as reaction medium for the synthesis of substituted polysaccharide derivatives [117,118]. ILs have proven great potential for large scale application due to their structural diversity, recyclability, high dissolution power, their high viscosity and hydrophilicity, and ability to be co-solvents in various systems [117,119–121]. Still, the information on wood surface modification in ILs is scarce, despite the high interest in cellulose functionalization in such medium.

Cellulose tosylation in ILs is usually performed in the presence of 1-allyl-3- methylimidazolium chloride (AMIMCl) or 1-butyl-3-methylimidazolium chloride (BMIMCl), at room temperature or below in order to avoid the tosyl reaction with chloride ions yielding in chlorodeoxycellulose as by-product, or the formation of insoluble cellulose derivatives by the cross-linking reaction between different hydroxyl groups [2,122]. However, cellulose tosylation in ILs at 10 °C yielded in a mixture of tosylated derivative (65%) and unmodified cellulose (45%) as result of the ineffective stirring of the reaction medium [117].

1-Butyl- and 1-benzylimidazole have been employed for homogeneous tosylation of cellulose when mixtures of IL and another co-solvent were used [123]. As both ILs have basicity comparable to imidazole and 1-methylimidazole but lower hydrophilic character, they can be easily extracted and recycled due to their high partition coefficient in the non-polar phase [117].

The synthesis of cellulose tosylates in homogeneous phase has been performed in IL/co-solvent mixtures where pyridine was employed, given its ability to act both as a base, by promoting the reaction, and as a co-solvent, reducing the viscosity of the medium during the entire reaction time [124,125].

In the presence of 1-ethyl-3-methylimidazolium acetate (EMIMAc), the cellulose tosylation evolved with the formation of cellulose acetates with high purity as result of the tosyl leaving group substitution by the acetate anions of the IL, evidenced by FTIR spectroscopy. Another possible pathway of the mechanism is the intermediary formation of a mixed anhydride by the reaction between tosyl chloride and EMIMAc, confirmed by NMR data [118,126]. Therefore, it is reasonable to assume that the formation of acetates during cellulose tosylation can be the result of both pathways: nucleophilic substitution of tosyl moiety and the reaction with the mixed toluenesulfonic-acetic anhydride formed in situ.

Special attention must be paid to the influence of the cation as it may affect the viscosity of the reaction mass. Thus, liquid imidazolium dialkyl phosphates cannot be used for the homogeneous tosylation at room temperature due to the rapid gelation of the reaction mixture upon the addition of the reagent, but the reaction is successful when tetraalkylammonium dialkyl phosphates were used under the same conditions [117,127].

Nevertheless, it is difficult to estimate the opportunity and feasibility of this method for the wood surface treatment, given the high capacity of ILs to dissolve and displace cellulose and, thus, to affect the wood structural integrity in some depth. Extensive research is still needed in this field as to design and test appropriate methodologies in order to apply this reaction to wood with satisfactory results. In addition, some studies evidenced that ILs are solvents not really as "green" as initially asserted, their reduced toxicity strongly depending on the structure, cation type, application, recovery and recycling, disposal etc. [117,128–132].

Aside from the immediate advantages of the tosylated wood, further chemical modification reactions can be performed on the same samples as the tosylation may be considered as a pre-treatment stage. Thus, the tosyl chloride is the activating agent as the tosyl moiety is the leaving group [95] which can be easily substituted by the direct reaction with various nucleophiles. These last mentioned can range from various carboxylic acids (employed in esterification reactions on wood, especially when acids with long aliphatic chains were used) [96] to carbanions or carboxylate anions (used for grafting some "living" polymer anions, such as polystyryl or polyacrylonitrile carbanions, on tosylated wood through an SN2 nucleophilic substitution) [97].

4. Wood Surface Modification by Esterification Reactions Using Carboxylic Acids

The subject has aroused the interest of the scientists for a long time [133] in order to identify methods to improve the properties of wood employed in various applications (mechanical and thermal properties, UV resistance, water uptake, biocide characteristics, etc.). Direct esterification reactions are, typically, difficult to perform due to their equilibrium which often leads to low yields, given the fact that the substrate (hydroxyl groups) is not readily accessible, and carboxylic acids have a lower reactivity as compared to their functional derivatives, such as anhydrides and acyl halides. Thus, the use of activating agents is a common practice even in the case of wood surface functionalization using carboxylic acids, and tosyl chloride was successfully used, as well as other activating agents such as N,N-dimethylacetamide (DMA), 1,1′-carbonyldiimidazole (CDA) or N,N-dicyclohexylcarbodiimide/4-pyrrolidinopyridine (DCC/PP).

4.1. *Wood Surface Modification by Esterification Reactions Using Fatty Acids*

Fatty acids are preferred for the esterification of wood under mild conditions as they are not aggressive toward the substrate, but, given their low reactivity, activation reagents or systems are required. The in situ activated wood esterification in the presence of TsCl was initially applied for acetylation [134] when a mixed anhydride is formed. The same mechanism occurred when carboxylic acids with long aliphatic chains were employed, ranging from 12 to 20 carbon atoms, yielding in a high degree functionalization of the available hydroxyl groups [96]. This method was further applied to the synthesis of the cellulose oxocarboxylic esters [135], and the resulted materials were stable at high temperature (above 300 °C).

For the case of long chain aliphatic carboxylic acids [116], namely caprylic (C_8), capric (or decanoic, C_{10}), lauric (C_{12}), palmitic (C_{16}) and, respectively, stearic (C_{18}) acid, the reaction medium consisted of DMA and LiCl, where DMA was solvent and base as well. The mechanism evolved through an intermediate stage when a mixed anhydride is formed by the reaction between TsCl and the carboxylic acid, as presented in Scheme 3. This sequence of reactions substantiates the molar ratio of reagents.

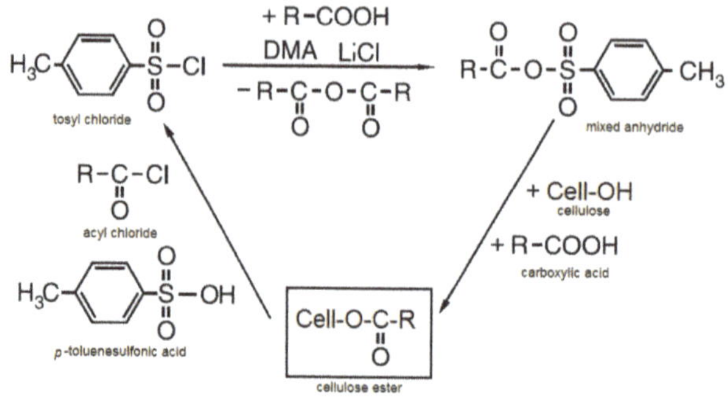

Scheme 3. Direct esterification of cellulose in the presence of TsCl and DMA/LiCl.

FTIR data confirmed the formation of the corresponding esters, and indicated no significant amounts of tosyl moieties as cellulose esters (as result of tosylation without further substitution) or as impurities, which is an indication that TsCl was acting as activating agent only. It was concluded that the degree of substitution as related to the available hydroxilic groups increased along with the number of carbon atoms of the acid when the acid and TsCl are in molar ratios corresponding to the mechanism (see Scheme 3) or higher, or when another base is added to the reaction medium, namely pyridine (Py). The thermal behavior of these esters follows the same pattern: the higher the number of carbon atoms in the aliphatic chain, the higher the degradation temperature, but all values remained in a narrow interval: 290–320 °C.

For comparison, the acetylation of cellulose has been performed with acetic anhydride using two different solvent systems: DMA/LiCl (a non-derivatizing solvent) and DMSO/TBAF (dimethyl sulfoxide/tetrabutyl ammonium fluoride trihydrate—a derivatizing solvent), at 40 °C, for 70 h. The results indicated lower values for the degree of substitution due to competing secondary reactions (formation of acetaldehyde as by-product, high rate of hydrolysis of the reagent) [116]. Nevertheless, this approach has proved to be effective for the case of various vinyl carboxylic acids, such as vinyl acetate, butyrate, laurate and, respectively, benzoate, in DMSO/TBAF medium, when high substitution degrees were achieved, but for long reaction time intervals as well (70 h).

A particular study case is the direct synthesis of adamantoyl cellulose by the direct reaction with the 1-adamantane carboxylic acid (Ad–COOH). Ad–COOH is a carboxylic acid with 11 carbon atoms, but it is not a fatty acid as it has a saturated cyclic structure; the presence of an adamantoyl moiety in the structure of various materials grants them biocidal activity [136].This direct esterification takes place in the presence of TsCl and DMA/LiCl/Py (Scheme 4), at 80 °C, for 24 h, or at room temperature for the same reaction time [7].

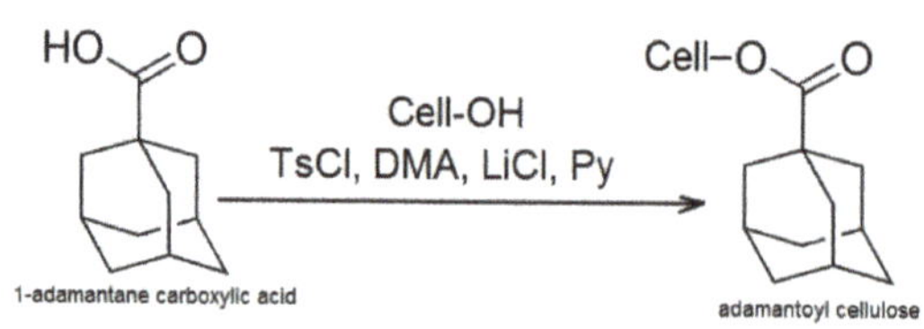

Scheme 4. Direct synthesis of adamantoyl cellulose.

Experimental data indicated this method afforded the highest degree of substitution as compared to the direct esterification in the presence of 1,1′-carbonyldiimidazole (CDA) (Scheme 5) or N,N-dicyclohexyl carbodiimide/4-pyrrolidinopyridine (DCC/PP) as activating agents [137,138].

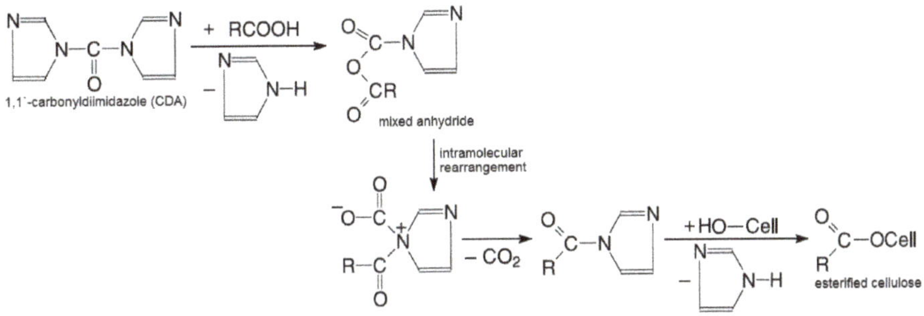

Scheme 5. Direct esterification of cellulose with carboxylic acids in the presenceof CDA as activating agent.

An interesting feature was recorded when CDA was employed: the esterification did not occur at room temperature, even when the molar ratios of Ad–COOH and CDA were higher than required by the mechanism, while at 80 °C the substitution degree was lower than for TsCl/DMA/LiCl. Nevertheless, the esterification using adamantoyl chloride (AdCl) in the presence of DMA/LiCl/Py [116,139] enabled most of the AdCl (85%) to react with the available hydroxylic groups in cellulose, which entailed degrees of substitution slightly higher than in the case of the activating system TsCl/DMA/LiCl.

Another very effective activating system for the esterification of wood using fatty acids was DMAc/LiCl and acetic anhydride as co-reagent [140]. The reaction took place at 130 °C, for 5 h, when –OH available groups were readily modified by the long-chain acyl cations by the means of a mixed anhydride formed in situ (as presented in Scheme 6).

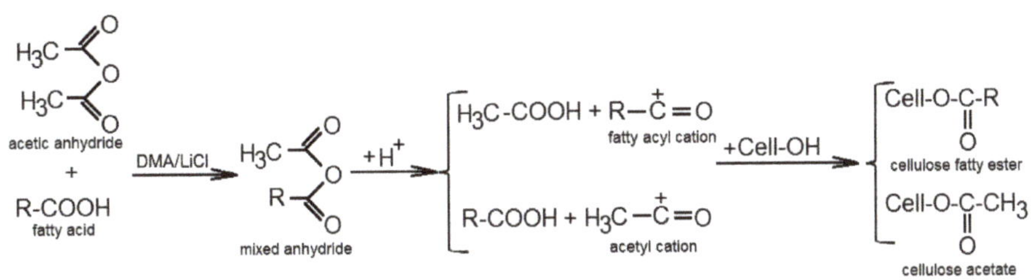

Scheme 6. Mechanism of direct esterification with fatty acids in the presenceof DMA/LiCl and acetic anhydride as co-reagent.

In the absence of acetic anhydride, the reaction performed under the same conditions yielded very low amounts of ester. The intermediate mixed anhydride may undergo a dissociation reaction due to the pKa discrepancy between the acetic acid and fatty acids which all have pKa values higher that of acetic acid. Therefore, the formation of the fatty acid cation (the alkanoyl group) is favored, but a significant amount of acetyl groups competes for the esterification as well, hence the mixed esters then resulted from the reaction. It was also possible to assess a correlation between the weight increase and degree of substitution in the fatty acid series n-octanoic-, n-decanoic-, n-dodecanoic-, n-tetradecanoic-, n-hexadecanoic- and n-octadecanoic acid, respectively. Thus, the highest weight increase was recorded for the acids with the highest molecular weight, provided that no significant degradation occurred as in other esterification systems [141].

4.2. *Wood Surface Modification by Esterification Reactions Using Unsaturated Carboxylic Acids*

The use of unsaturated carboxylic acids for the esterification of wood was motivated by the need to improve its dimensional stability which was one of the effects of the newly achieved reduced hygroscopicity. Studies have been conducted on beech wood samples submitted to direct esterification in the presence of trifluoroacetic anhydride (TFAA) which acted as activating agent [142,143]. The acylating species ($RC^{+}O$) resulted from the dissociation of the mixed anhydride obtained by the reaction of TFAA with the acid (acrylic, methacrylic, trans-crotonic, adipic, fumaric, maleic, mesaconic, citraconic, phthalic acid). The rate of this intermediate stage can be increased by using an acid catalyst, such as sulfuric acid, but its use is strongly limited by the degradative side effects on the wood support as it can dislodge lignin, cellobiose and other wood components [142,143]. Thus, the dissociation equilibrium depends on the pKa values of the selected acids as related to TFAA, and on the molecular weight of acids to a lesser extent. The IR data indicated the simultaneous formation of both carboxyl and trifluoroacetyl esters, and the latter are not easy to remove from the surface of modified wood by washing with water as it would be expected [142]. The same conclusions ensued for vinylacetic, tiglic, p-methylcrotonic, sorbic, 10-undecylenic, and elaidic acid, respectively [143]. Most of the selected acids yielded a weight increase after esterification of wood, but the weight decrease recorded after esterification with acrylic, methacrylic and elaidic acid indicated a degradation phenomenon associated with a loss of the structural integrity of wood, in spite of the significant decrease in the value of anti-shrinking efficiency.

The direct esterification of wood in the presence of TFAA was previously employed for the modification of beech (*Fagus sylvatica* L.) sawdust and sapwood from Ponderosa pine (Pinus ponderosa D.) samples with various acids [144]. The results were different: the esterification of the beech sawdust caused a discoloration of wood due to the degradation of trifluoroacetic esters in time, at room temperature, while the esterified Ponderosa pine chips acquired, in addition, a decrease in impact strength.

4.3. Wood Surface Modification by Esterification Reactions Using Polycarboxylic Acids

Polycarboxylic acids (PCAs) are reagents of high interest for the esterification of wood as surface treatment due to their structure and to the remarkable improvement in properties recorded for the modified wood, mainly durability and dimensional stability, but hygroscopicity and resistance toward biologic attack as well [145–148]. The most prominent representative of this class is citric acid (CA), but other acids were used as well, such as tartaric acid (TA), 1,2,3,4-butanetetracarboxylic acid (BTCA), and trimellitic acid (as anhydride which is more stable), although to a lower extent (Figure 4).

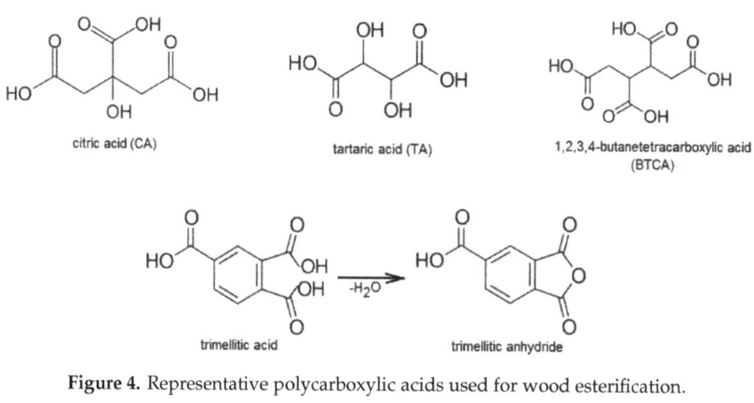

Figure 4. Representative polycarboxylic acids used for wood esterification.

All these multifunctional reagents are able to react with the available active moieties in the cell wall polymers [149]; still, BTCA is more effective in cross-linking than CA as it contents a higher number of –COOH groups, but CA is cheaper and easily accessible (either natural or synthetic) [150], as well as environmentally friendly [145,148]. The main characteristic of this method is the ability of these acids to form multiple ester bonds, which results in a tridimensional cross-linked network.

The mechanism evolves in two stages, as presented schematically in Scheme 7, when internal anhydrides between two adjacent COOH groups are successively formed and each of them reacts with units of cellulose from different sites.

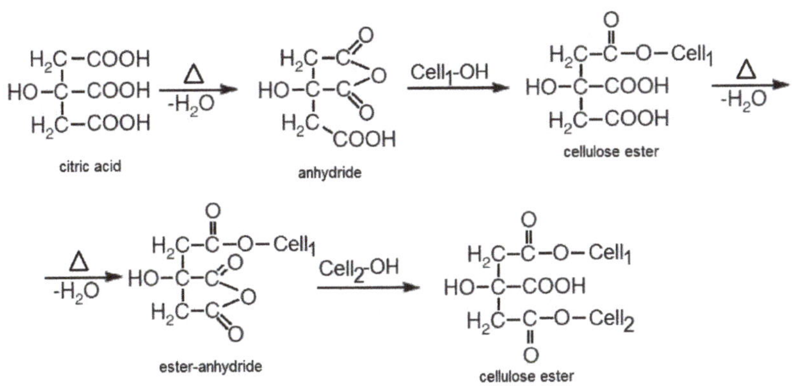

Scheme 7. Mechanism of esterification with citric acid (CA).

The second ester bond is usually formed when the reaction is performed either at high temperatures or at moderate temperatures but for longer time intervals. It is possible for the remaining carboxylic groups to interact with others in the wood structure and form a mixed anhydride which may further yield in another ester bond. In the case of BTCA, its ability to form esters is higher than that of CA, hence, the higher cross-linking density.

In most cases, the direct esterification required the use of a catalyst, such as sodium hypophosphite (NaHP), as well as a thermal treatment. In example, fir (Abies alba Mill) and beech (*Fagus sylvatica* L.) samples were treated with BTCA and CA, in the presence of NaHP, at 140 °C for 10 h, or by microwaves at 2.5 GHz, 750 W for 35 min [151]. The concentration of catalyst may also vary: 2.1% [151], or 6.5% [152,153], but the levels of performance of the esterified wood are in the same range. The effect of the concentration of NaHP on the properties of wood esterified with CA was studied over a wide range of concentration values, but at low temperatures in two stages (40 °C for 24 h, then 120 °C for 24 h) [154]. Three series of tests were designed, as follows: (a) 8% CA-1%, 3%, 5%, 10%, 15% NaHP; (b) 4%, 12%, 20%, 30% CA-without NaHP; and (c) 4%, 12%, 20%, 30% CA-5% NaHP. The characteristics used as comparison criteria were dimensional and weight stability of treated wood samples (anti-swelling efficiency—ASE), and mechanical properties (elasticity, impact strength, compression strength). The results were somewhat unexpected: the esterification occurred successfully in the absence of catalyst, but its presence caused slightly different effects on the properties of treated wood samples. Thus, the esterification without catalyst caused a weight increase in wood of 36% (when cell walls acquired a level of saturation), which entailed increased bulking (7%), ASE (50%), and compression strength (48%), but lower modulus of rupture (30%) and impact strength (50%). These effects can be explained by the higher density of the modified wood as compared to raw samples, which resulted in an enhanced stiffness. Therefore, NaHP is not indispensable if the reaction is allowed longer time intervals.

High compression strength values were also registered for fir (Abies alba Mill.) and beech (*Fagus sylvatica* L.) samples treated with CA by the microwave method [153], when polar macromolecules in wood (cellulose, lignin, hemicelluloses) underwent activation by polarization under the effect of the high energy field. It is well known that the microwaves treatment causes a rather homogeneous and fast heating in bulk, and the energy and mass flow have the same direction. As result, higher amounts of CA were linked to wood as compared to the typical thermal treatment, evidenced by the weight increase values, with the observation that fir wood retained a larger amount of CA than beech wood, but the values of the compression strength were lower, probably due to an inconsistent distribution of ester bonds in fir as effect of its low permeability. The same species submitted to the same treatment, but using BTCA instead CA, have showed a higher weight increase after esterification, which entailed reduced water sorption and improved ASE, and a higher cross-linking density due to the presence of 4 carboxylic moieties per each molecule of acid [151].

Remarkable biologic resistance also was achieved by esterification with CA. Thus, beech wood (*Fagus sylvatica* L.) samples were modified with CA/NaHP, for 10 h, at 140 °C, when an increase of 6.1% in wood weight was recorded and the resistance against brown rot fungi Poria placenta was over eight times higher than for raw wood [145]. For comparison, the same esterification performed on pine sapwood yielded in a weight increase of 12.4% and biological resistance against some rot fungi of 5.3 times higher biological durability [155].

For wood used in outdoors applications, it is of utmost importance to resist toward photochemical degradation processes that occur upon UV irradiation and are favored by other associated environmental factors, such as day/night or seasonal temperature variation, humidity, corrosion and exfoliation caused by high speed winds, biological attack [156,157]. This objective can be reached by modification of wood with polycarboxylic acids, namely CA. Experiments performed using 6.5% NaHP as catalyst have indicated that beech wood (*Fagus sylvatica* L.) samples underwent some color modification due to the esterification of cellulose rather than lignin, which is UV-sensitive [152]. During accelerated weathering tests run on both modified and unmodified samples for comparison, it was noticed that all samples showed similar cracking behavior. So, the cross-linking generated by CA was not effective enough to control and limit the degradation effects.

Wood modified by PCAs proved to have a valuable characteristic, namely the ability to retain various chemicals by surface adsorption, such as copper and lead ions upon esterification with CA [158,159], or cadmium, nickel and zinc when TA was employed [160]; aniline [161]; tetracycline [162]. For all cases, the esterification provided supplemental –COOH groups that remained un-reacted and available to retain various solutes from their corresponding aqueous solutions (according to the esterification mechanism presented in Scheme 8). For example, aspen wood samples treated with CA at 130 °C for various intervals (4 and 6 h, respectively) displayed high weight gains (52.55% and 58.43%, respectively) [159], and this enabled them to retain large amounts of Cu^{2+}, but depending on the pH of the medium which influences the acidic groups ionization and the mechanism of the metal ion removal. Better results would be expected in the case of wood modified with BTCA as it potentially introduces more available –COOH groups. In practice, it has been proven that it only caused an increased cross-linking density due to its enhanced ability to react with hydroxylic moieties in wood, and, thus, leaving less available adsorption sites [159]. Considering the pine (Pinus densiflora Siebold & Zucc.) sawdust modified with CA or TA, the maximum adsorption capacity was significantly higher than that of raw sawdust (14–57 times higher), and depended on the contact time and pH [160].

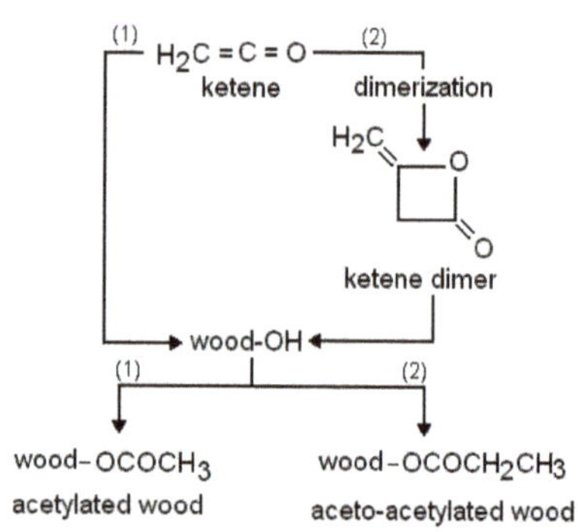

Scheme 8. Mechanism of wood esterification with ketene.

In the case of other chemicals, the adsorption mechanism is based on the formation of hydrogen bonds between the CA-modified wood and the adsorbed molecules. Thus, an effective adsorption of tetracycline from wastewater has been performed using wood modified with CA as sorbent [162]. The retention capacity of treated wood increased 8–12 times in comparison with raw wood, and the H-bonding formation massively occurred at pH = 5. When the wood samples were pre-treated with alkaline solutions (2% and 6%, respectively), the esterification was favored due to the fact that more cellulose units became available consequent to the lignin removal during the reaction of wood with NaOH. Thus, a higher number of carboxylic moieties can be active adsorption sites where H-bonds are formed between –COOH groups in esterified wood and O=C link in amide groups from zwitterionic tetracycline. This study exploited the low reactivity of lignin toward CA, so it was partially removed through the alkaline pre-treatment. In contrast, it has been substantiated that esterification reactions with CA took place with both lignin and cellulose, when some internal stabilization re-arrangements may occur in lignin structures [163]. These particular lignin structural rearrangements may contribute to the improved water resistance of the treated wood, as previously evidenced [164].

Other studies on the adsorption mechanism of CA-esterified wood have shown that the retention of aniline on modified wood (pine sawdust) was significantly enhance when β-cyclodextrin (β-CD) was used along with CA [161] due to a complex mechanism, when the additional hydrophobic interactions and the formation of inclusion complexes with β-CD came to increase the effect of H-bonding.

The ability of PCAs to bridge different wood components macromolecular chains through ester bonds and thus creating tridimensional cross-linked structures, given their multiple functional groups was investigated in order to promote modified wood as binder and/or adhesive. In example, CA-modified wood (poplar) samples were used to study the effect of this surface functionalization as promoter of bonding wood particle boards or other wood joints, or even wood veneer panels [164], as this behavior was already reported for other lignocellulosic materials [165–167]. In the case of PCA-modified wood, the binding effect is granted by the cumulative esterification of cellulose and lignin as well, despite the low reactivity of lignin and some local steric hindering due to the supramolecular architecture of wood components. It has been also concluded that CA used for the esterification of spruce samples [164] may have a strong catalytic effect which may be additionally responsible for the overall improved properties of the functionalized wood and, at the same time, favor the rearrangements of lignin at the modified surface. Thus, it was possible to exploit these findings and obtain flat wood pieces bonded by an active interface made of CA-modified wood (such as veneers, laminated veneer lumber—LVL—panels, plywood), having good mechanical properties in some cases (LVL panels showed increased shear strength as compared to standard, but the veneer-support binding failed upon bending) [163].

The same effect of reactive binding has been demonstrated on small wood particles (powder made of Acacia mangium Willd.) [168] processed by molding in the presence of CA, under pressure (4 MPa for 10 min) and at high temperature (140–220 °C). The mechanical characteristics of all samples were satisfactory, but particle boards manufactured at 180 °C showed excellent bending properties. The resistance toward water improved, especially at boiling, while thermal stability of materials enhanced in a direct relation with the processing temperature. These experimental results are due to the increased number of ester bonds formed at the surface of wood particles, confirmed by FTIR spectroscopy data, and this behavior can be explained by a combination of factors and their cumulative effects:

- the small size of wood particles created a significantly increased active surface for functionalization;
- the high amount of CA (20 wt.%) provided a larger number of carboxylic groups;
- the temperature regime favored the reaction rate and degree of substitution;
- the pressure applied during processing enhanced the penetration of CA molecules to reaction sites not easily accessible;
- the short heating time prevented further degradative thermo-chemical processes.

Other reaction systems based on PCAs were studied in relation with wood's functionalization, aiming to maximize the properties improvement in modified wood, limit the formation of by-products and waste (thus reducing the negative environmental impact), or even to employ reagents from renewable resources as measure of sustainable development (such as sorbitol and glycerol). Thus, CA is obtained from biomass, TA results as by-product from grape juice fermentation during wine production [147]. Other chemicals may also come from agro-industrial waste: sorbitol results from starch enzymatic hydrolysis of to dextrose followed by the catalytic hydrogenation of dextrose to sorbitol [169] or can be obtained from biomass by cellulose hydrolysis followed by glucose hydrogenation on mesoporous molecular sieves (porous silica and Ru-functionalized porous silica) [170], while glycerol yielded as by-product from the synthesis of bio-diesel [171].

Comparative studies on beech wood (*Fagus sylvatica* L.) samples were conducted using CA or TA, and mixtures of acids with glycerol (Gly) [147]. The concentration of acid solutions was 33% (TA) and 34% (CA), while the mixtures contained 10% Gly. After impregnation, the pressure (1 MPa, 60 min) and temperature (100–160 °C) regime was applied in order to maximize the treatment effects. However, the impregnation rate had comparable values for CA and CA + Gly, and TA and TA + Gly, respectively, although values for esterification with TA were slightly higher that for TA + Gly. The same conclusion is valid for the anti-swelling efficiency. The weight gain was not as significant as expected, either. It was also noticed that bending strength for samples treated with CA or TA strongly decreased with the increasing processing temperature, while samples treated with CA + Gly or TA + Gly behaved similar to raw samples, but better that acid-modified ones. This may be explained by the formation of the polyester matrix that allowed a high cross-linking density in the wood cell walls. Nevertheless, resistance toward decay [white rot fungi Trametes (Coriolus) versicolor)] was improved by using acid-Gly mixtures, with the observation that the best results for TA + Gly were obtained at the highest temperature, so, CA + Gly was more effective for the biological protection.

White pine (*Pinus strobus* L.) and lodgepole pine (*Pinus contorta* D.) specimens were treated with CA + Gly to improve their performance as siding materials in outdoor applications [148]. It was concluded that the treatment enhanced the dimensional stability, biological resistance and hardness as compared to untreated wood, although adhesion and density have not improved as expected, maybe due to a poor penetration of reagents. Further investigations on lodgepole pine (Pinus contorta D.) samples treated with CA + Gly from renewable resources, and intended for outdoor sidings [172], evidenced their improved properties (the service life expectancy for modified wood was estimated at 2.8 times longer than raw wood), but the comparative LCA studies on treated and raw samples indicated undesired environmental effects, where the most aggressive stage is the chemical treatment.

An optimization study on CA + Gly reaction system was performed on specimens of white pine (*Pinus strobus* L.) and lodgepole pine (*Pinus contorta* D.) [173] and the FTIR data confirmed a high density of ester bridges. At the same time, the cross-linking length and density caused increased thermal stability. The most effective catalysts were hydrochloric acid (HCl) and p-toluene sulfonic acid (p-TSA), but the amount of CA has also influenced the process. Another observation was that the deeper the reagent penetrated the wood substrate the more stable are the ester bonds and the higher the properties improvement.

Recently, sorbitol (S) has been considered and employed in wood esterification systems based on CA [146,174]. The modification treatment performed on the European beech wood (*Fagus sylvatica* L.) specimens with mixtures of CA and S in various ratios(10%, 20%, 30% and 50% *w*/*w*, respectively), and at different temperatures (140 and 160 °C, respectively), yielded materials with improved properties [146].Thus, optimum results were obtained for the solution with concentration 30% *w*/*w*, regardless of the temperature regime, when ASE, modulus of elasticity and decay resistance (tested against white-rot fungi [*Trametes (Coriolus) versicolor*], brown-rot fungi (*Coniphora puteana*), and soft rotting microfungi) has been increased. Despite these improvements, modulus of rupture and bending strength significantly decreased. Furthermore, studies on samples of pine sapwood (*Pinus sylvestris* L.) submitted to the same treatment [174] indicated that 140 °C is the temperature that allows the highest weight gain, although an optimization may be of interest. All samples have achieved an enhanced decay resistance to brown-rot fungi (*Postia placenta*) and white-rot fungi [*Trametes (Coriolus) versicolor*], as well as to blue-stain fungi (*Aureobasidium pullulans; Sydowia polyspora*; mixed culture of *A. pullulans, Cladosporium cladosporioides, Ulocladium atrum*, and *Sydowia polyspora*). The main issue of these materials is their leachability even after curing at high temperatures and long time (100–140 °C; 72 h) [147,174], which is an indication that the reagents adsorbed into the wood structure were not entirely linked by ester bonds to the wood cell walls constitutive polymers.

A rather recent and interesting experimental approach was the use of nano-sized particulate additives aiming at either an overall improvement in properties [175] or to expand the range of applications [176]. In example, nanoparticulate clay (Cloisite 30B) was employed as catalyst in a CA + Gly reaction system used for functionalization of lodgepole pine wood samples [175], and it proved to be an effective catalyst, yielding in materials with better properties than those obtained in the presence of HCl as catalyst. Dimensional stability and hardness were improved when Cloisite 30B was used and the thermal curing was performed at 180 °C, and the adhesion strength loss was lower. Nevertheless, given the following premises:

- the amount of nanoclay (2%) and its total exfoliation claimed by authors, and supported with XRD and TEM data, and the nanoclay structural features;
- HCl favored the multiple esterification, unlike Cloisite 30B, proved by thermal data;
- for samples prepared with nanoclay, density was not significantly improved compared to HCl samples, and the water vapor sorption was poor (this is an indication of the increased tortuosity of the material, which entailed slow transfer of water vapors in and out of the material), although dimensional stability and hardness increased;
- knowing the reinforcing effect of nanoclays in different types of composite materials, then it is possible to conclude that nanoparticulate clay may have been acting more as a reinforcing agent than a catalyst, even in low amounts. Further studies with nanoclay in various ratios are, though, necessary to elucidate the role of this component and its mechanism of action.

Multi-component systems based on wood sawdust (olive wood sawdust) and CA were designed in order to obtain materials containing magnetite, and with the ability to retain on their surface metal ions from wastewater [176]. Previous studies have confirmed already the ability of sawdust loaded with magnetic nanoparticles and CA-modified cellulose to retain metal ions from aqueous solutions [177,178]. Then, composites where CA was bond to either wood or magnetite particles were thus prepared and their study indicated that the CA-coated magnetite/wood was the material with a remarkable ability to retain metal ions (Cu^{2+}, Co^{2+} and Zn^{2+}) [176], as the process was fast and spontaneous, endothermic, and selective (adsorption of Zn^{2+} and Co^{2+} were significantly affected by the presence of Cu^{2+}) so far it could be used for analytic purposes.

4.4. Wood Surface Modification by Esterification Reactions Using Other Acids

It is not yet known exactly how much lignin and hemicelluloses in wood are able to react under these conditions, but experimental studies have attempted to provide a realistic parallel. Thus, the acetylation of sisal fibers conducted in different solvents has allowed researchers to conclude that the solvent system DMSO/TBAF is able to solubilize the cellulose and promote its functionalization with good results if only the water content of the reaction medium is reduced [135]. On the other hand, hemicelluloses from delignified poplar chips were successfully esterified in homogeneous media (DMF/LiCl and TEA), at temperatures in the range 45–75 °C, but using various acyl chlorides, when over 75% of the available –OH groups were stearoylated under optimum conditions [179]. Experiments conducted with other acids concluded that lignin and hemicelluloses from fast-growing poplar species can undergo both esterification and etherification reactions [180]. So, the 4-hydroxycinammic acid (HCA) esterified most of the available –OH groups in the lignin side chains, but 3-methoxy-4-hydroxycinnamic acid (ferulic acid, FA) was able to bridge lignin fragments by ester or ether bonds formed intra- or intermolecular. Even more, p-hydroxybenzoic, syringic and, respectively, vanillic acids were able to undergo esterification reactions mainly with lignin.

Benzoic acid and some of its derivatives, namely 2-nitro-, 3-nitro-, 4-nitrobenzoic acid, and 4-azidobenzoic acid, were employed for the direct esterification in the presence of pyridine and methane sulfonyl chloride as activating system [181]. A satisfactory degree of substitution (ranging from 0.9 for benzoic acid to 3.0 for 2-nitro- and 3-nitrobenzoic acids) was enabled by using a molar ratio acid:cellulose unit of 3.33:1 (taking into consideration the number of available –OH groups), and maintaining the reaction system at 50 °C for approx. 3–5 h. Still, it was not effective to extend the reaction time for the system with 4-nitrobenzoic and 4-azidobenzoic acid, as the degree of substitution did not exceed the values 2.0 and 1.0, respectively. This confirmed that the reaction rate and degree of substitution are strongly dependent on the acid reactivity, which is influenced by the nature and position on the aromatic ring of the electron attracting substituents.

In the case of esterification with benzoic acid, the formation of a mixed O-benzoyl-O-methane sulfonyl-cellulose ester was evidenced by IR spectroscopy confirming that methane sulfonyl groups were active enough to bind onto the glucopyranose unit. This mesylation of cellulose is a very slow reaction under the afore-mentioned conditions, but even so it could compete with the main esterification reaction.

Benzoylation of wood has been proved to be a very efficient method to protect wood against photochemical degradation caused by weathering and UV exposure. Experimental studies on benzoylated Scots pine wood samples confirmed the beneficial effects of this chemical modification [182]. Thus, the yield of the treatment was high as the weight gain of wood was approx. 70%–72%. It was demonstrated the benzoylation was significantly more effective in the photochemical stabilization of lignin than other chemical treatments (acetylation, alkylation), considering that the main stage in wood photochemical degradation is the absorption of UV radiation by lignin followed by its degradation, when various radicals are formed and, subsequently, they are acting as promoters of the photochemical degradation of cellulose and hemicelluloses [183,184]. It is not yet fully understood if this effect of reducing the amount of free radicals during weathering and UV exposure is due to the capacity of benzoyl moieties to absorb the UV radiation or to their ability to scavenge free radicals. Nevertheless, the study showed that the higher is the amount of benzoylated wood in the sample, the lower are the mass losses during the weathering tests.

Naphtenic acids were also employed for the protection of wood [185–187]. The nomenclature refers to all carboxylic acids present in crude oil. They are alkyl-substituted cycloaliphatic monocarboxylic acids, having the carboxylic group usually attached to the side chain than directly to the cycloalkane ring, and the naphtene moiety consists of cyclopentane and cyclohexane derivatives, although the mixtures of naphtenic acids may also contain low amounts of fatty acids, hydroxy and dibasic acids, phenolic compounds, sulfur compounds and water [188]. Wood modification with naphtenic acids is applied mainly using metal salts (sodium [185] or copper [186]), but organic derivatives (tributyltin naphtenate [187]) were also employed. Due to the chemical reactions that occurred in the wood, this treatment has proved to be effective against insect attack, weathering and UV radiation.

Other carboxylic acids were used too. Thus, cellulose formates were synthesized for analytical purposes by the direct reaction with formic acid for long intervals (4–15 days) or in the presence of an acidic catalyst (sulfuric acid) [189,190]. The one-pot esterification of wood and hydrolysis of modified cellulose was performed by treating softwood samples with molten oxalic acid dihydrate at 110 °C, for various time intervals (15–120 min), under reflux, when cellulose oxalates resulted [191,192]. This procedure allowed the preparation of cellulose nanocrystals in high yields (80%). But esterification of wood with oxalic acid was also employed to improve the moisture and water resistance of wood [193] or to enhance the interfacial interactions between the polymer matrix and wood reinforcements in wood-polymer composites [194].

Rosin is a mixture of extractives compounds originating in pine wood species that primarily consist of acids with various structures (resin acids 90%–95%), fatty acids included, phenolic compounds, terpenes and terpenoids. Commonly, resin acids are acid derivatives of diterpenes (abietane, labdane) and diterpenoids (pimarane and isopimarane) present in rosin in various amounts [195]. Abietic-type acids (abietic, neoabietic, palustric, levopimaric, and dehydroabietic acid) are predominant in rosin acids in comparison with pimarane-type acids (pimaric, isopimaric, and sandaracopimaric acid), but their ratios vary within large limits depending on the species and origin, time of harvesting, processing and storage, etc.

Wood treatment with rosin acids acquired new importance in recent years as it was employed in the surface modification of fast-growing wood species such as poplar (*Populus tormentosa* Carr. or *Populus* spp.) [196,197] aiming at improving the mechanical properties, dimensional stability, water resistance of wood samples, as well as their surface hardness. Experimental studies reported that wood modification was performed by impregnation followed by a thermal treatment. It is reasonable to assume that esterification reactions took place, given the processing conditions (solvent medium, temperature, pressure, long time intervals), although some specific esterification reactions of resin acids are conducted in industry at 260–300 °C, in the presence of metal oxides as catalysts [195]. This hypothesis is furthermore supported by the significant improvement in wood properties as it appears from the literature. In addition, the sizing process, well-known in paper industry, can be considered as another confirmation since it can be performed using not only rosin [198,199], but alkyl succinic anhydride (ASA) and alkyl ketene dimer (AKD) as well, reagents known to bind –OH groups in cellulose through ester bridges.

An interesting approach for wood modification is the use of amino acid tricine [74] which is N-[tri(hydroxymethyl)methyl] glycine, a zwitterionic amino acid, under the form of a crystalline powder moderately soluble in water. It is suitable for wood impregnation due to its reduced molecular size and its water solubility. Tricine can have beneficial effects after application on wood substrates through deposition of tricine crystals into wood cell walls when it further imparts increased mechanical strength (authors mentioned improved properties such as hardness and tensile strength for the tricine-modified wood), as well as by reaction with wood polymers, mainly polysaccharides, when cause a reduced tendency to retain moisture. As a zwitterion, tricine is involved in strong electrostatic and ionic interactions. For example, it can interact with carboxylate or phenolate moieties from wood polymer components through the presence of its structural amino function and is prone to induce formation of new hydrogen bonds with positive effect upon wood densification process. A previous work of the authors applying tricine for wood impregnation [75] has evidenced an increased wood UV resistance and a reduced degradation of lignin after such treatment.

4.5. Wood Surface Modification by Miscellaneous Procedures

4.5.1. Wood Esterification with Ketene

Esterification with ketene gas is an interesting approach due to the fact that it evolves without generating by-products (acetic acid). Basically, ketenes are simple organic compounds with the formula $R_1R_2C{=}C{=}O$ (where substituents R_1 and R_2 may be identic or different), highly reactive due to their particular chemical structure, and, therefore, are considered valuable reagents or intermediates in various organic syntheses. Since most of them are unstable, they are prepared *in situ* and used as they are formed [200]. Unfortunately, in the case of surface esterification of wood, the ketene gas has a poor ability to penetrate the wood in order to react with the hydroxyl groups in the cell walls [2,201]. Despite this drawback, it was possible to obtain southern pine and aspen wood flakes modified with ketene gas in a solvent-less procedure, when the reaction mixtures were maintained at 55 °C, for long reaction time intervals (10–15 h, as it was found that this esterification occurs with very low rates), in order to achieve weight gains of approx. 20% [202]. The surface reactions evolve through two mechanisms: (1) the direct reaction of ketene with –OH groups, when acetylated wood resulted, and (2) the dimerization of ketene under the reaction conditions, followed by the reaction of the dimer with the –OH moieties in the cell walls, yielding in aceto-acetylated wood. The mechanism is schematically presented in Scheme 8.

Some authors reported on the morphological changes occurred during the esterification with ketene gas and evidenced by scanning electron microscopy (SEM) [203,204]. Thus, cellulose Iα in wood transformed into cellulose IIβ which entailed the decrease of the fiber cross-section dimensions, and an enhanced smoothness of the fibers surface was noticeable.

At the same time, a remarkable reduction in water uptake (24.5%) was recorded for acetylated wood as compared to untreated samples. Furthermore, this treatment was also used for wood in order to obtain acetylated cellulose fibers after the removal of lignin and hemicelluloses [203] or dry and hydrophobic microfibrils of cellulose [205].

A more convenient variant of this method is using a stable compound, namely the alkyl ketene dimer (AKD) which is, basically, a 4-membered ring system of 2-oxetanone, having long alkyl chains (C_{12}–C_{22}) pendant to C3 and C4 [206]. By esterification of wood with AKD, β-keto-ester bridges are formed, and the alkyl moieties are transferred to the wood surface thus contributing to the hydrophobization of the treated wood samples (see Scheme 9). The mechanism may also include a concurrent side reaction when AKD reacts with water in the reaction system leading to the formation of the corresponding β-keto acid, which can spontaneously decompose to the corresponding ketone and carbon dioxide [207].

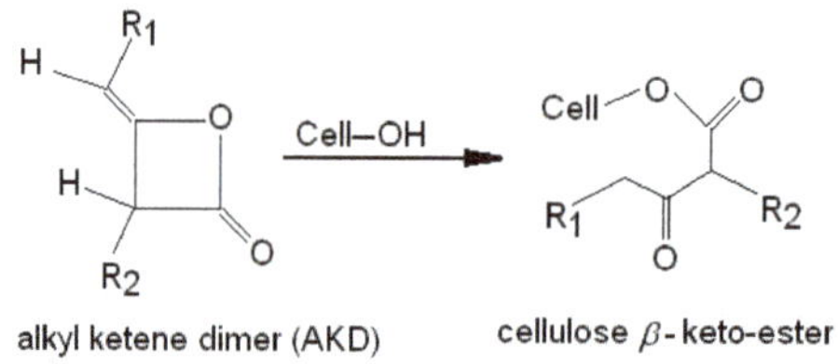

Scheme 9. Esterification of wood with alkyl ketene dimer (AKD).

The method was successfully applied to wood intended for further use in various composite formulations, such as: PP-wood fibers [28], plywood made of alder (*Alnus glutinosa* L. Gaertn.) and beech (Fagus orientalis Lipsky) wood logs [60], and particleboards bonded with urea-formaldehyde resin [62]. It can be concluded that AKD acted in the composites not only as an esterification reagent, but as a compatibilizating agent as well, due to its structure. The alkyl chains interpenetrated the polymer matrix macromolecules creating thus an inter-phase with increased thickness and well defined properties, given the fact that the adhesion and cohesion at the interface were provided by both physical and chemical bonding.

In addition to the remarkable decreased water absorption [62], the use of wood treated with AKD in composites caused a significant increase in the mechanical properties of composites. For samples of PP containing high amounts of poplar wood (70 wt.% wood), the tensile strength increased with 41%, modulus-45%, and the impact strength-38% [61]. Reduced water uptake of alder and beech plywoods was also associated with a considerably improved biologic resistance against white [*Trametes* (*Coriolus*) *versicolor*] and brown (Coniophera puteana) rot fungi [206].

4.5.2. Direct Esterification of Wood with ε-Caprolactone

A particular esterification applied for wood is the reaction with ε-caprolactone, in the presence of tin octanoate [$Sn(Oct)_2$] as initiator. The grafting onto wood cell walls takes place through ester bridges and the ring-opening polymerization of lactone (PCL) occurs under the processing conditions [208,209]. The impregnation of wood samples [Norway spruce Picea abies (L.) H. Karst.] with ε-caprolactone solution in DMF or toluene for 24 h prior to polymerization reaction (95 °C, 18 h) [208] is a key step in order to maximize the treatment effect. SEM micrographs evidenced a homogeneous distribution of the grafted PCL, although small amounts of unbound PCL remained inside cells after the filtration stage, and filled the micropores. The modified wood showed significantly improved mechanical properties, resistance toward water and dimensional stability: water repellence and dimensional stability increased by 50% and 45% respectively, when spruce, pine or poplar and other wood species were submitted to this particular esterification [208,210].

4.5.3. Transesterification Reactions for Wood Modification

Another method used for wood esterification is the transesterification in solvent medium (DMF, toluene), at temperature (90–120 °C), in the presence of catalyst (potassium carbonate), and using vinyl esters with various substituents, such as: vinyl ester of carboxyphenyl boronic acid [211], vinyl propionate, butyrate, pivalate, decanoate, stearate, crotonate, methacrylate, cinnamate, benzoate, and vinyl 4-tert-butylbenzoate [212,213]. The main feature of this approach is the continuous exchange of alkyloxy-groups at the surface of wood cell walls, and the continuous removal of the alcohol (or aldehyde) resulted as by-product allows the reaction equilibrium to be shifted toward satisfactory yields, knowing that esterification reactions typically evolve with low yields. It was noticed that the number of esterified –OH groups in wood decreased along with the increasing molecular weight of the reagent, which was confirmed by Fourier-transform infrared spectroscopy (FTIR) and ^{13}C cross-polarization with magic-angle spinning nuclear magnetic resonance spectroscopy (^{13}C CP–MAS NMR), probably due to a limited diffusion within the wood cell walls.

It was reported that poplar wood fibers modified by transesterification with vinyl acetate, propionate and benzoate [213] show increased thermal stability and attenuated hydrophilic character. The corresponding composite formulations with HDPE proved to be more stable under UV exposure, displayed improved mechanical properties and the weight losses caused by biological attack (brown rot and white rot fungi) were considerably reduced.

Another variant is the transesterification of acetylated wood (maritime pine sapwood *Pinus pinaster* Soland) with methyl benzoate, in the presence of dibutyltin oxide (DBTO) as a catalyst, and ethyl trimethylsilylacetate (ETMSA) [214], when it was evidenced that acetyl/benzoyl exchange rate increased along with the increasing amount of catalyst, temperature and reaction time, although a concurrent degradation reaction was also pointed out. Benzoylated wood proved to have an enhanced weathering and photochemical stability [215].

Silane derivatives were employed for wood transesterification, as well, in order for wood samples to acquire significantly pronounced hydrophobic character, and experimental data confirmed the replacement of acetyloxy groups and bonding of the silane moieties to wood [216–218]. Optimization studies indicated the yield of the transesterification depends on the reaction temperature and the presence of the catalyst (dibutyltin oxide).

Depending on the procedure, the properties of the treated wood were different. Thus, tetraethoxysilane (TEOS), methyl triethoxysilane (MTES) and propyl triethoxysilane (PTEO) were employed to modify wood samples of hinoki (Chamaecyparis obtuse Endl.), and two different experimental protocols were considered: (i) impregnation of wood with solutions of silane monomers (silane procedure), and (ii) impregnation with pre-hydrolysed oligomeric silanes (sol-gel process) [219]. It was therefore found that dimensional stability, moisture uptake and durability were significantly improved in wood treated with silane monomers (i), despite the fact that the weight gain was lower in comparison with the wood treated with silane oligomers (ii). On the other hand, for the wood samples esterified with silanes having larger alkyl groups, the improvement in properties was more pronounced.

An inexpensive and ecologically benign alternative is wood solvent-less transesterification with a water-based emulsion based on vegetable oil and resol [220], when the vegetable oil (consisting of fatty acids triglycerides) is the transesterification reagent, while the resol (a phenolic resin based on resorcinol-tannin-cashew nut shell liquid) was employed to increase the process effectiveness. After transesterification, the samples have displayed enhanced hydrophobicity and durability, but the improvement in their tensile strength depended on the type and time of treatment, as well as on the pH of the environment where composting experiments were conducted.

5. Conclusions

Wood represents a very complex and versatile natural material that can be modified in many ways to prepare it for direct use or even to activate its surface for further treatments. Modern treatments are physical (mainly plasma and densification procedures), chemical, and enzyme assisted or enzymatic grafting of different functional molecules to the wood surface, and application of thin films (coatings) and deposition of nano-sized particles by sol-gel techniques. Some of the surface modification methods have been introduced a long time ago and are still in use, other strategies of modification have been developed in the recent years.

All these pathways have the main goal to improve or even enlarge the range of properties of wood, such as good adhesion of glues and surface finishes, hydrophobic behavior and/or better hygroscopicity toward water and waterborne formulations. An improved resistance of wood surface against weathering processes when exposed to outdoor conditions is attained by grafting functional molecules onto cell walls and treatment with nanoparticles. Environmental benefits conferred by many wood surface modification technologies recently developed include reduced toxicity of the process, increased wood products life, and reduced maintenance cycles, lower negative impact on the environment and carbon footprint of the products. The resistance against wood decay microorganisms remains a significant topic for such investigations.

Nowadays, an interdisciplinary approach to increase the life of modified wood products used in outdoor and indoor applications (claddings or structural applications) is known as the "four Ds" rules of design and these are: deflection (shedding of water as defense way), drainage (fast removal of water as essential hazard), drying (restoring the optimal conditions as a prerequisite reaction) and durability (improved properties, mainly decay resistance, fire resistance).

As related to the wood surface modification strategies using esterification reactions, the use of natural compounds and their derivatives may represent a feasible trend that can be successfully implemented without major alterations of present technologies. Polysaccharides (such as starch, cellulose, hemicelluloses and lignin-carbohydrate complexes, maltodextrin), caffeine, propolis and other bee products derivatives, etc. are of high interest. The wise choice of treatments will remain a key step as more methods under development are progressing from laboratory scale to industrial scale production. Even combinations of different suitable methods (for example, applying a physical treatment prior to the esterification that would subsequently occur under mild conditions) are under research as they have the potential to maximize the overall effects.

Funding: This research received no external funding.

Acknowledgments: We dedicate this work to the memory of our colleague and friend, Ruxanda Bodîrlău (6 April 1953–26 June 2016), and thank her for her discretion, dedication and enthusiasm for as long as we have been working together.

Conflicts of Interest: The authors declare no conflict of interest.

References

1. Gérardin, P. New alternatives for wood preservation based on thermal and chemical modification of wood—A review. *Ann. For. Sci.* **2016**, *73*, 559–570. [CrossRef]
2. Rowell, R.M. Chemical modification of wood. In *Handbook of Wood Chemistry and Wood Composites*, 2nd ed.; Rowell, R.M., Ed.; Chapter 15; CRC Press: Boca Raton, FL, USA, 2012; pp. 537–598.
3. Rowell, R.M. Chemical modification of wood: A short review. *Wood Mat. Sci. Eng.* **2006**, *1*, 29–33. [CrossRef]
4. Hill, C.A.S. *Wood Modification: Chemical, Thermal and Other Processes*; John Wiley & Sons Ltd.: Chichester, UK, 2006.
5. Norimoto, M. Chemical modification of wood. In *Wood and Cellulose Chemistry*, 2nd ed.; Hon, D.N.S., Shiraishi, N., Eds.; Marcel Dekker: New York, NY, USA, 2001; pp. 573–598.

6. Kudanga, T.; Prasetyo, E.N.; Sipila, J.; Guebitz, G.M.; Nyanhongo, G.S. Reactivity of long chain alkylamines to lignin moieties: Implications on hydrophobicity of lignocellulose materials. *J. Biotechnol.* **2010**, *149*, 81–87. [CrossRef] [PubMed]
7. Suurnakki, A.; Buchert, J.; Gronqvist, S.; Mikkonen, H.; Peltonen, S.; Viikari, L. Bringing new properties to lignin rich fiber materials. *VTT Symposium* **2006**, *244*, 61–70.
8. Yamaguchi, H.; Maeda, Y.; Sakata, I. Bonding among woody fibers by use of enzymatic phenol dehydrogenative polymerization. *Mokuzai Gakkaishi* **1994**, *40*, 185–190.
9. Teacă, C.-A.; Roşu, D.; Mustată, F.; Rusu, T.; Roşu, L.; Roşca, I.; Varganici, C.-D. Natural bio-based products for wood coating and protection against degradation: A review. *BioResources* **2019**, *4*, 4873–4901.
10. Bodîrlău, R.; Spiridon, I.; Teacă, C.-A. Chemical investigation of wood tree species in a temperate forest, east-northern Romania. *BioResources* **2007**, *2*, 41–57.
11. Bodîrlău, R.; Teacă, C.-A.; Spiridon, I. Chemical modification of beech wood: Effect on thermal stability. *BioResources* **2008**, *3*, 789–800.
12. Zelinka, S.L.; Ringman, R.; Pilgård, A.; Thybring, E.E.; Jakes, J.E.; Richter, K. The role of chemical transport in the brown-rot decay resistance of modified wood. *Int. Wood Prod. J.* **2016**, *7*, 66–70. [CrossRef]
13. Nagarajappa, G.B.; Pandey, K.K. UV resistance and dimensional stability of wood modified with isopropenyl acetate. *J. Photochem. Photobiol. B Biol* **2016**, *155*, 20–27. [CrossRef]
14. Ringman, R.; Pilgard, A.; Brischke, C.; Richter, K. Mode of action of brow rot decay resistance in modified wood: A review. *Holzforschung* **2014**, *68*, 239–246. [CrossRef]
15. Thybring, E.E. The decay resistance of modified wood influenced by moisture exclusion and swelling reduction. *Int. Biodeter. Biodegrad.* **2013**, *82*, 87–95. [CrossRef]
16. Baur, S.I.; Easteal, A.J. Improved photoprotection of wood by chemical modification with silanes: NMR and ESR studies. *Polym. Adv. Technol.* **2013**, *24*, 97–103. [CrossRef]
17. Xiao, Z.F.; Xie, Y.J.; Mai, C. The fungal resistance of wood modified with glutaraldehyde. *Holzforschung* **2012**, *66*, 237–243. [CrossRef]
18. Papadopoulos, A.N. Chemical modification of pine wood with propionic anhydride: Effect on decay resistance and sorption of water vapour. *Bioresources* **2006**, *1*, 67–74.
19. Papadopoulos, A.N.; Hill, C.A.S. The biological effectiveness of wood modified with linear chain carboxylic acid anhydrides against *Coniophoraputeana*. *HolzRohWerkst* **2002**, *60*, 329–332.
20. Xiao, Z.; Xie, Y.; Adamopoulos, S.; Mai, C. Effects of chemical modification with glutaraldehyde on the weathering performance of Scots pine sapwood. *Wood Sci. Technol.* **2012**, *46*, 749–767. [CrossRef]
21. Roşu, D.; Teacă, C.-A.; Bodîrlău, R.; Roşu, L. FTIR and color change of the modified wood as a result of artificial light irradiation. *J. Photochem. Photobiol. B Biol.* **2010**, *99*, 144–149. [CrossRef]
22. Jebrane, M.; Sèbe, G.; Cullis, I.; Evans, P.D. Photostabilization of wood using aromatic vinyl esters. *Polym. Degrad. Stab.* **2009**, *94*, 151–157. [CrossRef]
23. Evans, P.D.; Owen, N.L.; Schmid, S.; Webster, R.D. Weathering and photostability of benzoylated wood. *Polym. Degrad. Stab.* **2002**, *76*, 291–303. [CrossRef]
24. Chang, S.T.; Chang, H.-T. Comparisons of the photostability of esterified wood. *Polym. Degrad. Stab.* **2001**, *71*, 261–266. [CrossRef]
25. Roşu, D.; Bodîrlău, R.; Teacă, C.-A.; Roşu, L.; Varganici, C.-D. Epoxy and succinic anhydride functionalized soybean oil for wood protection against UV light action. *J. Clean. Prod.* **2016**, *112*, 1175–1183. [CrossRef]
26. Roşu, L.; Varganici, C.D.; Mustaţă, F.; Rusu, T.; Roşu, D.; Roşca, I.; Tudorachi, N.; Teacă, C.-A. Enhancing the thermal and fungal resistance of wood treated with natural and synthetic derived epoxy resins. *ACS Sustain. Chem. Eng.* **2018**, *6*, 5470–5478. [CrossRef]
27. Bodîrlău, R.; Teacă, C.-A.; Spiridon, I. Preparation and characterization of composites comprising modified hardwood and wood polymers/poly(vinyl chloride). *BioResources* **2009**, *4*, 1285–1304.
28. Zhang, Y.; Xue, Y.; Toghiani, H.; Zhang, J.; Pittman, C.U. Modification of wood flour surfaces by esterification with acid chlorides: Use in HDPE/wood flour composites. *Compos. Interfaces* **2009**, *16*, 671–686. [CrossRef]
29. Ou, R.; Wang, Q.; Wolcott, M.P.; Sui, S.; Xie, Y. Rheological behavior and mechanical properties of wood flour/high density polyethylene blends: Effects of esterification of wood with citric acid. *Polym. Compos.* **2014**, *37*, 553–560. [CrossRef]
30. Adekunle, K.F. Surface treatments of natural fibres—A review: Part 1. *Open J. Polym. Chem.* **2015**, *5*, 41–46. [CrossRef]

31. Kalia, S.; Kaith, B.S.; Kaur, I. Pretreatments of natural fibers and their application as reinforcing material in polymer composites—A review. *Polym. Eng. Sci.* **2009**, *49*, 1253–1272. [CrossRef]
32. Vick, C.B.; Rowell, R.M. Adhesive bonding of acetylated wood. *Int. J. Adhesion Adhesives* **1990**, *10*, 263–272. [CrossRef]
33. Frihart, C.R.; Brandon, R.; Beecher, J.F.; Ibach, R.E. Adhesives for achieving durable bonds with acetylated wood. *Polymers* **2017**, *9*, 731. [CrossRef]
34. Teacă, C.-A.; Tanasă, F.; Zănoagă, M. Multi-component polymer systems comprising wood as bio-based component and thermoplastic polymer matrices—An overview. *BioResources* **2018**, *13*, 4728–4769. [CrossRef]
35. Riedl, B.; Angel, C.; Prégent, J.; Blanchet, P.; Stafford, L. Effect of wood surface modification by atmospheric-pressure plasma on waterborne coating adhesion. *BioResources* **2014**, *9*, 4908–4923. [CrossRef]
36. Wang, H.Y.; Du, G.B.; Li, Q.; Xu, R.Y.; Yuan, S.F. Bonding performance of wood treatment by oxygen and nitrogen cold plasma. *Appl. Mechanics Mater.* **2014**, *633–634*, 583–588. [CrossRef]
37. Acda, M.N.; Devera, E.E.; Cabangon, R.J.; Ramos, H.J. Effects of plasma modification on adhesion properties of wood. *Int. J. Adhesion Adhesives* **2012**, *32*, 70–75. [CrossRef]
38. Kim, B.-S.; Chun, B.-H.; Woo, I.L.; Hwang, B.-S. Effect of plasma treatment on the wood flour for wood flour/PP composites. *J. Thermoplast. Compos. Mater.* **2009**, *22*, 21–28. [CrossRef]
39. Podgorski, L.; Chevet, B.; Onic, L.; Merlin, A. Modification of wood wettability by plasma and corona treatments. *Int. J. Adhesion Adhesives* **2000**, *20*, 103–111. [CrossRef]
40. Pelaez-Samaniego, M.R.; Yadama, V.; Lowell, E.; Espinoza-Herrera, R. A review of wood thermal pretreatments to improve wood composite properties. *Wood Sci. Technol.* **2013**, *47*, 1285–1319. [CrossRef]
41. Borysiak, S. Influence of wood mercerization on the crystallization of polypropylene in wood/PP composites. *J. Therm. Anal. Calorim.* **2012**, *109*, 595–603. [CrossRef]
42. Lu, J.Z.; Wu, Q.; McNabb, H.S., Jr. Chemical coupling in wood fiber and polymer composites: A review of coupling agents and treatments. *Wood Fiber Sci.* **2000**, *32*, 88–104.
43. Bodîrlău, R.; Teacă, C.-A.; Resmeriţă, A.-M.; Spiridon, I. Investigation of structural and thermal properties of different wood species treated with toluene-2,4-diisocyanate. *Cell. Chem. Technol.* **2012**, *46*, 381–387.
44. Maldas, D.; Kokta, B.V. Surface modification of wood fibers using maleic anhydride and isocyanate as coating components and their performance in polystyrene composites. *J. Adhes. Sci. Technol.* **1991**, *5*, 727–740. [CrossRef]
45. Li, Q.; Matuana, L.M. Surface of cellulosic materials modified with functionalized polyethylene coupling agents. *J. Appl. Polym. Sci.* **2003**, *88*, 278–286. [CrossRef]
46. Farsi, M. Wood–plastic composites: Influence of wood flour chemical modification on the mechanical performance. *J. Reinf. Plast. Compos.* **2010**, *29*, 3587–3592. [CrossRef]
47. Xie, H.; Jarvi, P.; Karesoja, M.; King, A.; Kilpelainen, I.; Argyropoulos, D.S. Highly compatible wood thermoplastic composites from lignocellulosic material modified in ionic liquids: Preparation and thermal properties. *J. Appl. Polym. Sci.* **2009**, *111*, 2468–2476. [CrossRef]
48. Teacă, C.-A.; Bodîrlău, R.; Spiridon, I. Maleic anhydride treatment of softwood—effect on wood structure and properties. *Cell. Chem. Technol.* **2014**, *48*, 863–868.
49. Bodîrlău, R.; Teacă, C.-A. Fourier transform infrared spectroscopy and thermal analysis of lignocellulose fillers treated with organic anhydrides. *Rom. J. Phys.* **2009**, *54*, 93–104.
50. Bodîrlău, R.; Teacă, C.-A. Softwood chemical modification by reaction with organic anhydrides. *Rev. Roum. Chim.* **2008**, *53*, 1059–1064.
51. Bodîrlău, R.; Teacă, C.-A.; Roşu, D.; Roşu, L.; Varganici, C.-D.; Coroabă, A. Physico-chemical properties investigation of softwood surface after treatment with organic anhydride. *Cent. Eur. J. Chem.* **2013**, *11*, 2098–2106.
52. Doczekalska, B.; Bartkowiak, M.; Zakrzewski, R. Modification of sawdust from pine and beech wood with the succinic anhydride. *HolzRohWerkst* **2007**, *65*, 187–191. [CrossRef]
53. Ichazo, M.N.; Albano, C.; Gonzalez, J.; Perera, R.; Candal, M.V. Polypropylene/wood flour composites: Treatments and properties. *Compos. Struct.* **2001**, *54*, 207–214. [CrossRef]
54. Yang, T.; Cao, J.; Ma, E. How does delignification influence the furfurylation of wood? *Ind. Crops Prod.* **2019**, *135*, 91–98. [CrossRef]

55. Sejati, P.S.; Imbert, A.; Gérardin-Charbonnier, C.; Dumarçay, S.; Fredon, E.; Masson, E.; Nandika, D.; Priadi, T.; Gérardin, P. Tartaric acid catalyzed furfurylation of beech wood. *Wood Sci. Technol.* **2017**, *51*, 379–394. [CrossRef]
56. Li, W.; Ren, D.; Zhang, X.; Wang, H.; Yu, Y. The furfurylation of wood: A nanomechanical study of modified wood cells. *BioResources* **2016**, *11*, 3614–3625. [CrossRef]
57. Li, W.; Wang, H.; Ren, D.; Yu, Y.S.; Yu, Y. Wood modification with furfuryl alcohol catalysed by a new composite acidic catalyst. *Wood Sci. Technol.* **2015**, *49*, 845–856. [CrossRef]
58. Pfriem, A.; Dietrich, T.; Buchelt, B. Furfuryl alcohol impregnation for improved plasticization and fixation during the densification of wood. *Holzforschung* **2012**, *66*, 215–218. [CrossRef]
59. Esteves, B.; Nunes, L.; Pereira, H. Properties of furfurylated wood (*Pinus pinaster*). *Eur. J. Wood Prod.* **2011**, *69*, 521–525. [CrossRef]
60. Gaye KöseDemirel, G.; Güdül, H.; Temiz, A.; Kuştaş, S.; Aydin, İ. Effect of alkyl ketene dimer on the physical, mechanical, and biological durability of plywood. *BioResources* **2017**, *13*, 147–156.
61. Zhang, H. Effect of a novel coupling agent, alkyl ketene dimer, on the mechanical properties of wood-plastic composites. *Mater. Design* **2014**, *59*, 130–134. [CrossRef]
62. Hundhausen, U.; Militz, H.; Mai, C. Use of alkyl ketene dimer (AKD) for surface modification of particleboard chips. *Eur. J. Wood Prod.* **2009**, *67*, 37–45. [CrossRef]
63. Rowell, R.M.; Wang, R.H.S.; Hyatt, J.A. Flakeboards made from aspen and southern pine wood flakes reacted with gaseous ketene. *J. Wood Chem. Technol.* **1986**, *6*, 449–471. [CrossRef]
64. Çetin, N.S.; Ozmen, N. Dimensional changes in Corsican and Scots pine sapwood due to reaction with crotonic anhydride. *Wood Sci. Technol.* **2001**, *35*, 257–267. [CrossRef]
65. Hill, C.A.S.; Cetin, N.S. Surface activation of wood for graft polymerization. *Int. J. Adhesion Adhesives* **2000**, *20*, 71–76. [CrossRef]
66. Del Menezzi, C.; Amirou, S.; Pizzi, A.; Xi, X.; Delmotte, L. Reactions with wood carbohydrates and lignin of citric acid as a bond promoter ofwood veneer panels. *Polymers* **2018**, *10*, 833. [CrossRef] [PubMed]
67. Amirou, S.; Pizzi, A.; Delmotte, L. Citric acid as waterproofing additive in butt joints linear wood welding. *Eur. J. Wood Prod.* **2017**, *75*, 651–654. [CrossRef]
68. Essoua Essoua, G.G.; Blanchet, P.; Landry, V.; Beauregard, R. Pine wood treated with a citric acid and glycerol mixture: Biomaterial performance improved by a bio-byproduct. *BioResources* **2016**, *11*, 3049–3072. [CrossRef]
69. Umemura, K.; Ueda, T.; Kawai, S. Characterization of wood-based molding bonded with citric acid. *J. Wood Sci.* **2012**, *58*, 38–45. [CrossRef]
70. Umemura, K.; Kawai, S. Development of wood-based materials bonded with citric acid. *For. Prod. J.* **2015**, *65*, 38–42. [CrossRef]
71. Feng, X.; Xiao, Z.; Sui, S.; Wang, Q.; Xie, Y. Esterification of wood with citric acid: The catalytic effects of sodium hypophosphite (SHP). *Holzforschung* **2014**, *68*, 427–433. [CrossRef]
72. Umemura, K.; Ueda, T.; Munawar, S.S.; Kawai, S. Application of citric acid as natural adhesive for wood. *J. Appl. Polym. Sci.* **2012**, *123*, 1991–1996. [CrossRef]
73. Bischof Vukusic, S.; Katovic, D.; Schramm, C.; Trajkovic, J.; Sefc, B. Polycarboxylic acids as non-formaldehyde anti-swelling agents for wood. *Holzforschung* **2006**, *60*, 439–444. [CrossRef]
74. Hauptmann, M.; Gindl-Altmutter, W.; Hansmann, C.; Bacher, M.; Rosenau, T.; Liebner, F.; D'Amico, S.; Schwanninger, M. Wood modification with tricine. *Holzforschung* **2015**, *69*, 985–991. [CrossRef]
75. Hauptmann, M.; Rosenau, T.; Gindl-Altmutter, W.; Hansmann, C. Effects of UV-irradiation on tricine impregnated wood. *Eur. J. Wood Prod.* **2014**, *72*, 617–622. [CrossRef]
76. Larsson-Brelid, P. *Benchmarking and State of-the-Art Report for Modified Wood*; SP Report no. 54; SP Technical Research Institute of Sweden: Stockholm, Sweden, 2013; pp. 1–31.
77. Militz, H. Treatment of timber with water soluble dimethylol resins to improve their dimensional stability and durability. *Wood Sci. Technol.* **1993**, *27*, 347–355. [CrossRef]
78. Xie, Y.; Krause, A.; Mai, C.; Militz, H.; Richter, K.; Urban, K.; Evans, P.D. Weathering of wood modifiedwith the N-methylol compound 1, 3-dimethylol-4, 5-dihydroxyethyleneurea. *Polym. Degrad. Stabil.* **2005**, *89*, 189–199. [CrossRef]

79. Xie, Y.; Krause, A.; Militz, H.; Turkulin, H.; Richter, K.; Mai, C. Effect of treatments with 1,3-dimethylol-4, 5-dihydroxyethyleneurea (DMDHEU) on the tensile properties of wood. *Holzforschung* **2007**, *61*, 43–50. [CrossRef]
80. Xie, Y.; Krause, A.; Militz, H.; Mai, C. Weathering of uncoated and coated wood treated withmethylated 1, 3-dimethylol-4, 5-dihydroxyethyleneurea (mDMDHEU). *Holz Roh Werkst.* **2008**, *66*, 455–464. [CrossRef]
81. Shin, Y.; Winder, E.M.; Han, K.S.; Lee, H.; Bonheyo, G.T. Enhanced capacities of mixed fatty acid-modified sawdust aggregators for remediation of crude oil spill. *ACS Omega* **2019**, *4*, 412–420. [CrossRef]
82. Salla, J.; Pandey, K.K.; Prakash, G.K.; Mahadevan, K.M. Photobleaching and dimensional stability of rubberwood esterified by fatty acid chlorides. *J. Wood. Chem. Technol.* **2012**, *32*, 121–136. [CrossRef]
83. Prakash, G.K.; Mahadevan, K.M. Enhancing the properties of wood through chemical modification with palmitoyl chloride. *Appl. Surf. Sci.* **2008**, *254*, 1751–1756. [CrossRef]
84. Wang, X.J.; Zhang, M.H.; Wang, X.M. Properties of esterificated wood with oxalic acid and cetyl alcohol. *Adv. Mater. Res.* **2010**, *123–125*, 1187–1190. [CrossRef]
85. Matsuda, H. Chemical modification of solid wood. In *Chemical Modification of Lignocellulosic Materials*; Hon, D.N.S., Ed.; Marcel Dekker: New York, NY, USA, 1996; pp. 159–183.
86. Nunez, A.J.; Kenny, J.M.; Reboredo, M.M.; Aranguren, M.I.; Marcovich, N.E. Thermal and dynamic mechanical characterization of polypropylene-wood flour composites. *Polym. Eng. Sci.* **2002**, *42*, 733–742. [CrossRef]
87. Cofta, G.; Borysiak, S.; Doczekalska, B.; Garbarczyk, J. Resistance of polypropylene-wood composites to fungi. *Polimery* **2006**, *51*, 276–279. [CrossRef]
88. Chang, H.-T.; Chang, S.-T. Modification of wood with isopropyl glycidyl ether and its effects on decay resistance and light stability. *Bioresour. Technol.* **2006**, *97*, 1265–1271. [CrossRef] [PubMed]
89. Hill, C.A.S.; Cetin, N.S.; Ozmen, Z. Potential catalysts for the acetylation of wood. *Holzforschung* **2000**, *54*, 269–272. [CrossRef]
90. Evans, P.D.; Wallis, A.F.A.; Owen, N.L. Weathering of chemically modified wood surfaces—Natural weathering of Scots pine acetylated to different weight gains. *Wood Sci. Technol.* **2000**, *34*, 151–165. [CrossRef]
91. Clemons, C.; Young, R.A.; Rowell, R.M. Moisture sorption properties of composite boards from esterified aspen fiber. *Wood Fiber Sci.* **1992**, *24*, 353–363.
92. Mishra, S.; Naik, J.B. Absorption of water at ambient temperature and steam in wood–polymer composites prepared from agro-waste and polystyrene. *J. Appl. Polym. Sci.* **1998**, *68*, 681–686. [CrossRef]
93. Iwamoto, Y.; Itoh, T. Vapor phase reaction of *wood* with maleic anhydride (I): Dimensional stability and durability of treated wood. *J. Wood Sci.* **2005**, *51*, 595–600. [CrossRef]
94. Nenkova, S.; Dobrilova, T.Z.V.; Natov, M.; Vasilieva, S.; Velev, P. Modification of wood flour with maleic anhydride for manufacture of wood-polymer composites. *Polym. Polym. Compos.* **2006**, *14*, 185–194. [CrossRef]
95. Tanasa, F. Hi-tech polymers by direct polycondensation reactions. In *Functional Polymeric Materials Designed for Hi-Tech Applications*; Nechifor, M., Ed.; Transworld Research Network: Kerala, India, 2010; pp. 155–175.
96. Sealey, J.; Samaranayake, G.; Todd, J.; Glasser, W.G. Novel cellulose derivatives. IV. Preparation and thermal analysis of waxy ester of cellulose. *J. Polym. Sci.* **1996**, *34*, 1613–1620. [CrossRef]
97. Narayan, R.; Biermann, C.J.; Hunt, M.O.; Horn, D.P. Cellulose graft copolymers for potential adhesive applications. In *Bonding of Plastics to Wood, volume 385, Adhesives from Renewable Resources*; ACS Symposium Series; Hemingway, R.W., Conner, A.H., Branham, S.J., Eds.; Chapter 24; ACS: Washington, DC, USA, 1989; pp. 337–354.
98. Rowell, R.M. Distribution of reacted chemicals in southern pine modified with acetic anhydride. *Wood Sci.* **1982**, *15*, 172–182.
99. Rowell, R.M. *Chemical Modification of Wood: A Review*; Commonwealth Forestry Bureau: Oxford, UK, 1983; Volume 6, pp. 363–382.
100. Mantanis, G.I. Chemical modification of wood by acetylation or furfurylation: A review of the present scaled-up technologies. *BioResources* **2017**, *12*, 4478–4489. [CrossRef]
101. Tarkow, H. *Decay Resistance of Acetylated Balsa*; USDA Forest Service, Forest Prod. Lab.: Madison, WI, USA, 1945; p. 4.
102. Rowell, R.M. Acetylation of wood—A review. *Int. J. Lignocellulosic Products* **2014**, *1*, 1–27.

103. Rowell, R. Acetylation of wood—Journey from analytical technique to commercial reality. *Forest Prod. J.* **2006**, *56*, 4–12.
104. Tarkow, H.; Stamm, A.J.; Erickson, E. *Acetylated Wood*; Report no. 1593; USDA Forest Service, Forest Products Laboratory: Madison, WI, USA, 1946; p. 29.
105. Rowell, R.M. The chemistry of solid wood. In *Advances in Chemistry Series*; No. 207; American Chemical Society: Washington, DC, USA, 1984; p. 540. [CrossRef]
106. Jones, D.; Sandberg, D.; Goli, G.; Todaro, L. *Wood Modification in Europe: A State-of-the Art about Processes, Products and Applications*; Firenze University Press: Firenze, Italy, 2019; p. 123.
107. Hill, C.A.S. Wood modification: An update. *BioResources* **2011**, *6*, 918–919.
108. Rahn, K.; Diamantoglou, M.; Klemm, D.; Berghmans, H.; Heinze, T. Homogeneous synthesis of cellulose p-toluenesulfonates in N,N-dimethylacetamide/LiCl solvent system. *Angew. Makromol. Chem. Appl. Macromol. Chem. Phys.* **1996**, *238*, 143–163. [CrossRef]
109. Ermeydan, M.A.; Cabane, E.; Gierlinger, N.; Koetz, J.; Burgert, I. Improvement of wood material properties via in situ polymerization of styrene into tosylated cell walls. *RSC Adv.* **2014**, *4*, 12981–12988. [CrossRef]
110. Ermeydan, M.A.; Cabane, E.; Masic, A.; Koetz, J.; Burgert, I. Flavonoid insertion into cell walls improves wood properties. *ACS Appl. Mater. Interfaces* **2012**, *4*, 5782–5789. [CrossRef]
111. Heinze, T.; Liebert, T. Unconventional methods in cellulose functionalization. *Prog. Polym. Sci.* **2001**, *26*, 1689–1762. [CrossRef]
112. Ma, P.; Gao, Y.; Zhai, H. Fractionated wheat straw lignin and its application as antioxidant. *BioResources* **2013**, *8*, 5581–5595. [CrossRef]
113. Ujihara, M.; Nakatsubo, F.; Katahira, R. A novel selective cleavage method for β-O-4 substructure in lignins named TIZ method. I. Degradation of guaiacyl and syringyl models. *J. Wood Chem. Technol.* **2003**, *23*, 71–87. [CrossRef]
114. Diop, A.; Awada, H.; Zerrouki, R.; Daneault, C.; Montplaisir, D. Tosylation and characterization of lignin in water. *Ind. Eng. Chem. Res.* **2014**, *53*, 16771–16776. [CrossRef]
115. Panovic, I.; Montgomery, J.R.; Lancefield, C.S.; Puri, D.; Lebl, T.; Westwood, N.J. Grafting of technical lignins through regioselective triazole formation on β-O-4 linkages. *ACS Sustain. Chem. Eng.* **2017**, *5*, 10640–10648. [CrossRef]
116. Heinze, T.; Liebert, T.F.; Pfeiffer, K.S.; Hussain, M.A. Unconventional cellulose esters: Synthesis, characterization and structure–property relations. *Cellulose* **2003**, *10*, 283–296. [CrossRef]
117. Ziemann, E. Surface modification of cellulose membranes. In *Reviews in Global Water Cycle*; OSF Preprints: Leeuwarden, The Netherlands, 2017. [CrossRef]
118. Kishida, A.; Mishima, K.; Corretge, E.; Konishi, H.; Ikada, Y. Interactions of poly(ethyleneglycol)-grafted cellulose membranes with proteins and platelets. *Biomaterials* **1992**, *13*, 113–118. [CrossRef]
119. Gardner, D.J.; Schulz, T.P.; McGinnis, G.D. The pyrolytic behavior of selected lignin preparations. *J. Wood Chem. Technol.* **1985**, *5*, 85–110. [CrossRef]
120. Gericke, M.; Fardim, P.; Heinze, T. Ionic liquids—Promising but challenging solvents for homogeneous derivatization of cellulose. *Molecules* **2012**, *17*, 7458–7502. [CrossRef]
121. Liebert, T.; Heinze, T. Interaction of ionic liquids with polysaccharides. 5. Solvents and reaction media for the modification of cellulose. *BioResources* **2008**, *3*, 576–601.
122. Heinze, T.; Gericke, M. Ionic liquids as solvents for homogeneous derivatization of cellulose: Challenges and opportunities. In *Production of Biofuels and Chemicals with Ionic Liquids*; Fang, Z., Smith, R.L., Jr., Qi, X., Eds.; Springer: Dordrecht, The Netherlands, 2014; pp. 107–144.
123. Vitz, J.; Erdmenger, T.; Haensch, C.; Schubert, U.S. Extended dissolution studies of cellulose in imidazolium based ionic liquids. *Green Chem.* **2009**, *11*, 417–424. [CrossRef]
124. Trivedi, P.; Fardim, P. Recent advances in cellulose chemistry and potential applications. In *Production of Materials from Sustainable Biomass Resources*; Fang, Z., Smith, R.L., Jr., Tian, X., Eds.; Springer: Singapore, 2019; pp. 99–115.
125. Granström, M.; Kavakka, J.; King, A.; Majoinen, J.; Mäkelä, V.; Helaja, J.; Hietala, S.; Virtanen, T.; Maunu, S.L.; Argyropoulos, D.S.; et al. Tosylation and acylation of cellulose in 1-allyl-3-methylimidazolium chloride. *Cellulose* **2008**, *15*, 481–488. [CrossRef]
126. Gericke, M.; Schaller, J.; Liebert, T.; Fardim, P.; Meister, F.; Heinze, T. Studies on the tosylation of cellulose in mixtures of ionic liquids and a co-solvent. *Carbohydr. Polym.* **2012**, *89*, 526–536. [CrossRef] [PubMed]

127. Möllmann, E.; Heinze, T.; Liebert, T.; Köhler, S. Homogeneous Synthesis of Cellulose Ethers in Ionic Liquids. U.S. Patent Application 20090221813 A1, 3 September 2009.
128. Vitz, J.; Yevlampieva, N.P.; Rjumtsev, E.; Schubert, U.S. Cellulose molecular properties in 1-alkyl-3-methylimidazolium-based ionic liquid mixtures with pyridine. *Carbohydr. Polym.* **2010**, *82*, 1046–1053. [CrossRef]
129. Kohler, S.; Heinze, T. Efficient synthesis of cellulose furoates in 1-N-butyl-3-methylimidazolium chloride. *Cellulose* **2007**, *14*, 489–495. [CrossRef]
130. Buchanan, C.M.; Buchanan, N.L.; Guzman-Morales, E. Cellulose solutions comprising tetraalkylammoniumalkyl phosphate and products produced therefrom. U.S. Patent Application WO/2010/120269, US2009/004626, 21 October 2010.
131. Jessop, P.G. Searching for green solvents. *Green Chem.* **2011**, *13*, 1391–1398. [CrossRef]
132. Deetlefs, M.; Seddon, K.R. Assessing the greenness of some typical laboratory ionic liquid preparations. *Green Chem.* **2010**, *12*, 17–30. [CrossRef]
133. Shimizu, Y.; Hayashi, J. Acylation of cellulose with carboxylic acids. *Cell. Chem. Technol.* **1989**, *23*, 661–670.
134. Shimizu, Y.; Hayashi, J. A new method for cellulose acetylation with acetic acid. *Sen'IGakkaishi* **1988**, *44*, 451–456. [CrossRef]
135. Heinze, T.; Schaller, J. New water soluble cellulose ester synthesized by an effective acylation procedure. *Macromol. Chem. Phys.* **2000**, *201*, 1214–1218. [CrossRef]
136. Orzeszko, A.; Gralewska, R.; Starosciak, B.J.; Kazimierczuk, Z. Synthesis and antimicrobial activity of new adamantane derivatives. *Acta Biochim. Pol.* **2000**, *47*, 87–94. [CrossRef]
137. Samaranayake, G.; Glasser, W.G. Cellulose derivatives with low DS. 1. A novel acylation system. *Carbohydr. Polym.* **1993**, *22*, 1–7. [CrossRef]
138. Samaranayake, G.; Glasser, W.G. Cellulose derivatives with low DS. 2. Analysis of alkanoates. *Carbohydr. Polym.* **1993**, *22*, 79–86. [CrossRef]
139. Gräbner, D.; Liebert, T.; Heinze, T. Synthesis of novel adamantoyl cellulose using differently activated carboxylic acid derivatives. *Cellulose* **2002**, *9*, 193–201. [CrossRef]
140. Vaca-Garcia, C.; Thiebaud, S.; Borredon, M.E.; Gozzelino, G. Cellulose esterification with fatty acids and acetic anhydride in lithium chloride/N, N-dimethylacetamide medium. *J. Am. Oil Chem. Soc.* **1998**, *75*, 315–319. [CrossRef]
141. Thiebaud, S.; Borredon, M.E. Solvent-free wood esterification with fatty acid chlorides. *Bioresour. Technol.* **1995**, *52*, 169–173. [CrossRef]
142. Nakagami, T.; Amimoto, H.; Yokota, T. Esterification of wood with unsaturated carboxylic acids. I.: Preparation of several wood-esters by the TFAA method. *Bull. Kyoto Univ. For.* **1974**, *46*, 217–224.
143. Nakagami, T.; Yokota, T. Esterification of wood with unsaturated carboxylic acids. II: Reaction conditions of esterification and properties of the prepared esters of wood. *Bull. Kyoto Univ. For.* **1975**, *47*, 178–183.
144. Arni, P.C.; Gray, J.D.; Scougall, R.K. Chemical modification of wood. I Use of trifluoroacetic anhydride in the esterification of wood by carboxylic acids. *J. Appl. Chem.* **1961**, *11*, 157–163. [CrossRef]
145. Despot, R.; Hasan, M.; Jug, M.; Šefc, B. Biological durability of wood modified by citric acid. *Drv. Ind.* **2008**, *59*, 55–59.
146. Mubarok, M.; Militz, H.; Dumarçay, S.; Gérardin, P. Beech wood modification based on in situ esterification with sorbitol and citric acid. *Wood Sci. Technol.* **2020**, *54*, 1–24. [CrossRef]
147. L'Hostis, C.; Thévenon, M.F.; Fredon, E.; Gérardin, P. Improvement of beech wood properties by in situ formation of polyesters of citric and tartaric acid in combination with glycerol. *Holzforschung* **2018**, *72*, 291–299. [CrossRef]
148. Essoua Essoua, G.G. Développement d'une Stratégie de Modification du bois afin de Limiter les Variations Dimensionnelles du Produit Lambris dans un Contexte Éco-Responsible. Available online: https://corpus.ulaval.ca/jspui/handle/20.500.11794/27111 (accessed on 10 June 2020).
149. Zoldners, J.; Kiseleva, T. Modification of hemicelluloses with polycarboxylic acids. *Holzforschung* **2013**, *67*, 567–571. [CrossRef]
150. Guo, W.; Xiao, Z.; Wentzel, M.; Emmerich, L.; Xie, Y.; Militz, H. Modification of Scots pine with activated glucose and citric acid: Physical and mechanical properties. *BioResources* **2019**, *14*, 3445–3458.
151. BischofVukusic, S.; Katovic, D.; Grgac, S.F.; Trajkovic, J.; Sefc, B.; Voncina, B. Study of the wood modification process with polycarboxylic acids and microwave treatment. *Wood Res.-Slovakia* **2010**, *55*, 121–130.

152. Miklečić, J.; Jirouš-Rajković, V. Accelerated weathering of coated and uncoated beech wood modified with citric acid. *Wood Ind./Drv.* **2011**, *62*, 277–282. [CrossRef]
153. Šefc, B.; Trajković, J.; Sinković, T.; Hasan, M.; Ištok, I. Compression strength of fir and beech wood modified by citric acid. *Wood Ind./Drv.* **2012**, *63*, 45–50. [CrossRef]
154. Berube, M.A.; Schorr, D.; Ball, R.J.; Landry, V.; Blanchet, P. Determination of in situ esterification parameters of citric acid-glycerol based polymers for wood impregnation. *J. Polym. Environ.* **2018**, *26*, 970–979. [CrossRef]
155. Hasan, M.; Despot, R.; Trajkovic, J.; Šefc, B. Role of modification processes at increasing biological durability of wood. In Proceedings of the International Conference European Union–Challenges and Perspectives for the Wood-Processing Industry, Zagreb, Croatia, 13 October 2006; pp. 153–158.
156. Teacă, C.A.; Bodîrlău, R. Photochemical behavior of wood-based materials. In *Photochemical Behavior of Multicomponent Polymeric-Based Materials*; Rosu, D., Visakh, P.M., Eds.; Springer International Publishing: Cham, Switzerland, 2016; pp. 91–107. [CrossRef]
157. Kranitz, K.; Sonderegger, W.; Bues, C.T.; Niemz, P. Effects of aging on wood: A literature review. *Wood Sci. Technol.* **2016**, *50*, 7–22. [CrossRef]
158. Low, K.S.; Lee, C.K.; Mak, S.M. Sorption of copper and lead by citric acid modified wood. *Wood Sci. Technol.* **2004**, *38*, 629–640. [CrossRef]
159. McSweeny, J.D.; Rowell, R.M.; Min, S.H. Effect of citric acid modification of aspen wood on sorption of copper ion. *J. Nat. Fibers* **2006**, *3*, 43–58. [CrossRef]
160. Yang, J.S.; Park, Y.T.; Baek, K.; Choi, J. Removal of metal ions from aqueous solutions using sawdust modified with citric acid or tartaric acid. *Sep. Sci. Technol.* **2010**, *45*, 1963–1974. [CrossRef]
161. Zhou, Y.; Gu, X.; Zhang, R.; Lu, J. Removal of aniline from aqueous solution using pine sawdust modified with citric acid and β-cyclodextrin. *Ind. Eng. Chem. Res.* **2014**, *53*, 887–894. [CrossRef]
162. Liu, M.; Zhao, Z.; Yu, W. Citric acid modified wood membranes for efficient adsorption of tetracycline: Effect of alkali pretreatment concentration and adsorption mechanism. *Chem. Eng. J.* **2020**, *393*, 124748. [CrossRef]
163. Widyorini, R.; Yudha, A.P.; Adifandi, Y.; Umemura, K.; Kawai, S. Characteristic of bamboo particleboard bonded with citric acid. *Wood Res. J.* **2013**, *4*, 31–35.
164. Kusumah, S.S.; Umemura, K.; Guswenrivo, I.; Yoshimura, T.; Kanajama, K. Utilization of sweet sorghum bagasse and citric acid for manufacturing of particleboard II: Influences of pressing temperature and time on particleboard properties. *J. Wood Sci.* **2017**, *63*, 161–172. [CrossRef]
165. Kusumah, S.S.; Umemura, K.; Oshioka, K.Y.; Miyafuji, H.; Kanajama, K. Utilisation of sweet sorghum bagasseand citric acid for manufacturing particleboard I: Effects of pre-drying treatment and citric acid content onthe board properties. *Ind. Crops Prod.* **2016**, *84*, 34–42. [CrossRef]
166. Liao, R.; Xu, J.; Umemura, K. Low density sugarcane bagasse particleboard bonded with citric acid and sucrose: Effect of board density and additive content. *BioResources* **2016**, *11*, 2174–2185. [CrossRef]
167. Widyorini, R.; Nugraha, P.; Rahman, M.; Prayitno, T. Bonding ability of a new adhesive composed of citric acid-sucrose for particleboard. *BioResources* **2016**, *11*, 4526–4535. [CrossRef]
168. Umemura, K.; Ueda, T.; Kawai, S. Effects of moulding temperature on the physical properties of wood-based moulding bonded with citric acid. *Forest Prod. J.* **2012**, *62*, 63–68. [CrossRef]
169. Young, N.W.G.; O'Sullivan, G.R. The influence of ingredients on product stability and shelf life. In *Food and Beverage Stability and Shelf Life*; Woodhead Publishing Series in Food Science, Technology and Nutrition; Woodhead Publishing: Cambridge, UK, 2011; pp. 132–183.
170. Romero, A.; Esther, E.; Sastre, Á.; Nieto-Márquez, A. Conversion of biomass into sorbitol: Cellulose hydrolysis on MCM-48 and d-Glucose hydrogenation on Ru/MCM-48. *Micropor. Mesopor. Mater.* **2016**, *224*, 1–8. [CrossRef]
171. Okoye, P.U.; Hameed, B.H. Review on recent progressin catalytic carboxylation and acetylation of glycerol as abyproduct of biodiesel production. *Renew. Sustain. Energy Rev.* **2016**, *53*, 558–574. [CrossRef]
172. Essoua Essoua, G.G.; Beauregard, R.; Amor, B.; Blanchet, P.; Landry, V. Evaluation of environmental impacts of citric acid and glycerol outdoor softwood treatment: Case-study. *J. Clean. Prod.* **2017**, *164*, 1507–1518. [CrossRef]
173. L'Hostis, C.; Fredon, E.; Thévenon, M.-F.; Santiago-Medina, F.-J.; Gérardin, P. Beech wood treated with polyglycerol succinate: A new effective method for its protection and stabilization. *Holzforschung* **2020**, *74*, 351–361. [CrossRef]

174. Larnøy, E.; Karaca, A.; Gobakken, L.R.; Hill, C.A.S. Polyesterification of wood using sorbitol and citric acid under aqueous conditions. *Int. Wood Prod. J.* **2018**, *9*, 66–73. [CrossRef]
175. Schorr, D.; Blanchet, P.; Essoua Essoua, G.G. Glycerol and citric acid treatment of lodgepole pine. *J. Wood Chem. Technol.* **2018**, *38*, 123–136. [CrossRef]
176. El-Sheikh, A.H.; Fasfous, I.I.; Al-Salamin, R.M.; Newman, A.P. Immobilization of citric acid and magnetite on sawdust for competitive adsorption and extraction of metal ions from environmental waters. *J. Environ. Chem. Eng.* **2018**, *6*, 5186–5195. [CrossRef]
177. Shah, J.; Jan, M.; Khan, M.; Amir, S. Removal and recovery of cadmium from aqueous solutions using magnetic nanoparticle-modified sawdust: Kinetics and adsorption isotherm studies. *Desalin. Water Treat.* **2015**, *57*, 9736–9744. [CrossRef]
178. Kuo, C.; Wu, C.; Chen, M. Adsorption of lead ions from aqueous solutions by citric acid-modified celluloses. *Desalin. Water Treat.* **2015**, *55*, 1–7. [CrossRef]
179. Sun, R.; Fanga, J.M.; Tomkinson, J.; Hill, C.A.S. Esterification of hemicelluloses from poplar chips in homogenous solution of N,N-dimethylformamide/lithium chloride. *J. Wood Chem. Technol.* **1999**, *19*, 287–306. [CrossRef]
180. Sun, R.C.; Sun, X.F.; Wang, S.Q.; Zhu, W.; Wang, X.Y. Ester and ether linkages between hydroxycinnamic acids and lignins from wheat, rice, rye, and barley straws, maize stems, and fast-growing poplar wood. *Ind. Crops Prod.* **2002**, *15*, 179–188. [CrossRef]
181. Talaba, P.; Srokova, I.; Hodul, P.; Ebringerova, A. New procedure for the preparation of cellulose esters with aromatic carboxylic acids. *Chem. Papers* **1996**, *50*, 365–368.
182. Hon, D.N.-S.; Ou, N.-H. Thermoplasticization of wood. I. Benzylation of wood. *J. Polym. Sci. Part A Polym. Chem.* **1989**, *27*, 2457–2482. [CrossRef]
183. Teacă, C.-A.; Roşu, D.; Bodîrlău, R.; Roşu, L. Structural changes in wood under artificial UV light irradiation determined by FTIR spectroscopy and color measurements—A brief review. *BioResources* **2013**, *8*, 1478–1507. [CrossRef]
184. Zanoaga, M.; Tanasa, F. Photochemical behavior of synthetic polymeric multicomponent materials composites and nanocomposites. In *Photochemical Behavior of Multicomponent Polymeric-Based Materials*; Rosu, D., Visakh, P.M., Eds.; Advanced Structured Materials, Springer: Cham, Switzerland, 2016; Volume 26, pp. 109–164.
185. Kevrešan, S.; Kovačević, B.; Ćirin-Novta, V.; Kuhajda, K.; Kandrač, J.; Pavlović, K.; Grbović, L. Biochemical changes in cuttings of *Robiniapseudoacacia* after treatment with naphthenate. *J. Serb. Chem. Soc.* **2007**, *72*, 953–959. [CrossRef]
186. Muhammed, S.; Musgrave, O.C.; Petty, J.A. Impregnation of rubberwood and other Malaysian timbers with copper naphthenate and trimethyl borate. *J. Trop. For. Sci.* **2009**, *21*, 345–352.
187. Dawson, B.S.W.; Kroese, H.W.; Hong, S.O.; Lane, G.T. Resin bleed after painting from radiata pine boards treated with tributyltin naphthenate (light organic solvent preservative) or copper, chromium and arsenic compounds (water-borne preservative). *HolzRohWerkstg* **2002**, *60*, 8–24. [CrossRef]
188. Brient, J.A.; Wessner, P.J.; Doyle, M.N. Naphthenic acids. In *Kirk-OthmerEncyclopaedia of Chemical Technology*, 4th ed.; Kroschwitz, J.I., Ed.; John Wiley and Sons: New York, NY, USA, 1995; pp. 1017–1029.
189. Schnabelrauch, M.; Vogt, S.; Klemm, D.; Nehls, I.; Philipp, B. Readily hydrolyzable cellulose esters as intermediates for the regioselective derivatization of cellulose, 1. Synthesis and characterization of soluble, low-substituted cellulose formates. *Angew. Makromol. Chem. Appl. Macromol. Chem. Phys.* **1992**, *198*, 155–164. [CrossRef]
190. Liebert, T.; Klemm, D.; Heinze, T. Synthesis and carboxymethylation of organo-soluble trifluoroacetates and formates of cellulose. *J. Macromol. Sci. Part A Pure Appl. Chem.* **1996**, *33*, 613–626. [CrossRef]
191. Li, D.; Henschen, J.; Ek, M. Esterification and hydrolysis of cellulose using oxalic acid dihydrate in a solvent-free reaction suitable for preparation of surface-functionalised cellulose nanocrystals with high yield. *Green Chem.* **2017**, *19*, 5564–5567. [CrossRef]
192. Jia, C.; Chen, L.; Shao, Z.; Agarwal, U.P.; Hu, L.; Zhu, J.Y. Using a fully recyclable dicarboxylic acid for producing dispersible and thermally stable cellulose nanomaterials from different cellulosic sources. *Cellulose* **2017**, *24*, 2483–2498. [CrossRef]
193. Akhtar, M.; Kenealy, W.; Horn, E.; Swaney, R.; Winandy, J. Medium-Density Fiberboard (MDF) with Improved Water Resistance. U.S. Patent No. 8,123,904 B2, 28 February 2012.

194. Gardea-Hernández, G.; Ibarra-Gómez, R.; Flores-Gallardo, S.G.; Hernández-Escobar, C.A.; Pérez-Romo, P.; Zaragoza-Contreras, E.A. Fast wood fiber esterification. I. Reaction with oxalic acid and cetyl alcohol. *Carbohydr. Polym.* **2008**, *71*, 1–8. [CrossRef]
195. Abdel-Raouf, M.E.S.; Abdul-Raheim, A.R.M. Rosin: Chemistry, Derivatives, and Applications: A Review. Available online: https://bioaccent.org/chemistry/chemistry39.php (accessed on 10 June 2020).
196. Dong, Y.; Yan, Y.; Wang, K.; Li, J.; Zhang, S.; Xia, C.; Shi, S.Q.; Cai, L. Improvement of water resistance, dimensional stability, and mechanical properties of poplar wood by rosin impregnation. *Eur. J. Wood Wood Prod.* **2016**, *74*, 177–184. [CrossRef]
197. Dong, Y.; Zhang, W.; Hughes, M.; Wu, M.; Zhang, S.; Li, J. Various polymeric monomers derived from renewable rosin for the modification of fast-growing poplar wood. *Compos. Part. B Eng.* **2019**, *174*, 106902. [CrossRef]
198. Wang, F.; Tanaka, H. Aminated poly-N-vinylformamide as a modern retention aid of alkaline paper sizing with acid rosin sizes. *J. Appl. Polym. Sci.* **2000**, *78*, 1805–1810. [CrossRef]
199. Matsushita, Y.; Iwatsuki, A.; Yasuda, S. Application of cationic polymer prepared from sulfuric acid lignin as a retention aid for usual rosin sizes to neutral papermaking. *J. Wood Sci.* **2004**, *50*, 540–544. [CrossRef]
200. Miller, R.; Abaecherli, C.; Said, A.; Jackson, B. Ketenes. In *Ullmann's Encyclopedia of Industrial Chemistry*, 6th ed.; Wiley-VCH: Weinheim, Germany, 2003; Volume 18, pp. 717–732.
201. Rowell, R.M. Chemical modification of wood. In *Handbook of Engineering Biopolymers, Homopolymers, Blends, and Composites*; Hanser Gardner Publications: Cincinnati, OH, USA, 2007; pp. 673–691.
202. Nilsson, T.; Rowell, R.M.; Simonson, R.; Tillman, A.-M. Fungal resistance of pine particle boards made from various types of acetylated chips. *Holzforschung* **1988**, *42*, 123–126. [CrossRef]
203. Azeh, Y.; Olatunji, G.A.; Mamza, P.A. Scanning electron microscopy and kinetic studies of ketene-acetylated wood/cellulose high-density polyethylene blends. *Int. J. Carbohydr. Chem.* **2012**, *7*. [CrossRef]
204. Azeh, Y.; Olatunji, G.A.; Oladoye, S.; Atolani, O. Ketene acetylated wood cellulose for industrial applications in wood-base and polymer industry. *J. Environ. Sci. Technol.* **2012**, *5*, 168–176. [CrossRef]
205. Yan, Y.; Amer, H.; Rosenau, T.; Zollfrank, C.; Dörrstein, J.; Jobst, C.; Zimmermann, T.; Keckes, J.; Veigel, S.; Gindl-Altmutter, W.; et al. Dry, hydrophobic microfibrillated cellulose powder obtained in a simple procedure using alkyl ketene dimer. *Cellulose* **2016**, *23*, 1189–1197. [CrossRef]
206. Mini-Encyclopedia of Papermaking Wet-End Chemistry. Available online: https://projects.ncsu.edu/project/hubbepaperchem/AKD.html (accessed on 10 June 2020).
207. Lindström, T.; Larsson, P.T. Alkyl ketene dimer (AKD) sizing—A review. *Nordic Pulp Paper Res. J.* **2008**, *23*, 202–209. [CrossRef]
208. Ermeydan, M.A.; Cabane, E.; Hass, P.; Koetz, J.; Burgert, I. Fully biodegradable modification of wood for improvement of dimensional stability and water absorption properties by poly (ε-caprolactone) grafting into the cell walls. *Green Chem.* **2014**, *16*, 3313–3321. [CrossRef]
209. Ermeydan, M.A.; Tomak, E.D.; Kartal, Z.N. Wood property improvement of siberian pine by combination of boric acid impregnation and in-situ polymerization of ε-caprolactone. *PoliteknikDergisi* **2019**, *22*, 157–161.
210. Ermeydan, M.A.; Gonultas, O.; Candan, Z. Chemical modification of paulownia, poplar, and eucalyptus wood by ε-caprolactone grafting inside cell walls to improve wood properties. *İleriTeknolojiBilimleriDergisi* **2017**, *6*, 323–330.
211. Jebrane, M.; Heinmaa, I. Covalent fixation of boron in wood through transesterification with vinyl ester of carboxyphenylboronic acid. *Holzforschung* **2016**, *70*, 577–583. [CrossRef]
212. Jebrane, M.; Sèbe, G. A new process for the esterification of wood by reaction with vinyl esters. *Carbohydr. Polym.* **2008**, *72*, 657–663. [CrossRef]
213. Wei, L.; McDonald, A.G.; Freitag, C.; Morrell, J.J. Effects of wood fiber esterification on properties, weatherability and biodurability of wood plastic composites. *Polymer Degrad. Stabil.* **2013**, *98*, 1348–1361. [CrossRef]
214. Özmen, N.; Çetin, N.S.; Tingaut, P.; Sèbe, G. A new route for the functionalisation of wood through transesterification reactions. *Eur. Polym. J.* **2006**, *42*, 1617–1624. [CrossRef]
215. Pandey, K.K.; Chandrashekar, N. Photostability of wood surfaces esterified by benzoyl chloride. *J. Appl. Polym. Sci.* **2005**, *99*, 2367–2374. [CrossRef]
216. Çetin, N.S.; Özmen, N.; Tingaut, P.; Sèbe, G. New transesterification reaction between acetylated wood and tetramethoxysilane: A feasibility study. *Eur. Polym. J.* **2005**, *41*, 2704–2710. [CrossRef]

217. Özmen, N.; Çetin, N.S.; Tingaut, P.; Sèbe, G. Transesterification reaction between acetylated wood and trialkoxysilane coupling agents. *J. Appl. Polym. Sci.* **2007**, *105*, 570–575. [CrossRef]
218. Pries, M. Treatment of Solid Wood with Silanes, Polydimethylsiloxanes and Silica Sols. Available online: https://d-nb.info/106277065X/34 (accessed on 12 June 2020).
219. Donath, S.; Militz, H.; Mai, C. Wood modification with alkoxysilanes. *Wood Sci. Technol.* **2004**, *38*, 555–566. [CrossRef]
220. Saha, P.; Manna, S.; Sen, R.; Roy, D.; Adhikari, B. Durability of lignocellulosic fibers treated with vegetable oil–phenolic resin. *Carbohydr. Polym.* **2012**, *87*, 1628–1636. [CrossRef]

coatings

MDPI

Article

Paint Pull-Off Strength and Permeability in Nanosilver-Impregnated and Heat-Treated Beech Wood

Hamid Taghiyari [1,*], Ayoub Esmailpour [2] and Antonios Papadopoulos [3,*]

1 Wood Science and Technology Department, Faculty of Materials Engineering & New Technologies, Shahid Rajaee Teacher Training University, Tehran 14115, Iran
2 Department of Physics, Faculty of Sciences, Shahid Rajaee Teacher Training University, Tehran 14115, Iran; esmailpour@sru.ac.ir
3 Laboratory of Wood Chemistry and Technology, Department of Forestry and Natural Environment, International Hellenic University, GR-6 6 00 Drama, Greece
* Correspondence: htaghiyari@sru.ac.ir (H.T.); antpap@teiemt.gr (A.P.)

Received: 7 October 2019; Accepted: 30 October 2019; Published: 1 November 2019

Abstract: The effects of impregnation with nanosilver suspension as well as heat treatment on pull-off adhesion strength and specific air permeability in beech specimens were studied here. The size range of silver nanoparticles was 30–80 nm. The cross-section of specimens was cold-sprayed with unpigmented sealer-clear, polyester, and lacquer paints. Heat treatment, as the most commonly used wood modification, was applied at three different temperatures of 145, 165, and 185 °C. Results showed that the highest and lowest pull-off strengths were found in the un-impregnated and unheated specimens painted with polyester (8.98 MPa) and the unpainted unheated nanosilver-impregnated specimens (3.10 MPa), respectively. Impregnation with nanosilver resulted in the rupture of perforation plates and pit openings, and eventually, permeability increased significantly. As for the pull-off adhesion strength, the increased permeability resulted in the adhesive being penetrated in to the pores in the wood substrate, and eventually, a significant decrease in the pull-off strengths occurred. No significant correlation was found between pull-off strength versus specific air permeability, although both properties depend on the porous structure. This was due to the fact that permeability depends on the continuous pore system, while pull-off strength is dependent on the surface pore system of the substrate.

Keywords: coating; modification; nanotechnology; un-pigmented paints; permeability; pull-off

1. Introduction

As solid woods are natural porous media, many of their properties depend on the size of the pores, the way they are interconnected or isolated from the neighboring pores, and even the quality of the surface of the pores [1–4]. Many factors influence the formation of wood and its structure and porous system thereof—factors such as initial spacing, intercropping with different plants, drying procedures [5–7], growing season, extractive content, moisture content, and hygroscopicity of wood [8,9]. Therefore, researchers constantly look for new methods to modify wood for better mechanical properties because few species offer radial and axial uniformity in their produced wood [10].

Thermal modification is generally considered the most commercially-used wood modification method [11]. It has been recognized as a method to improve the dimensional stabilization of wood and increase its decay resistance [11–13]. Thermal modification at high temperatures has decreasing effects on some mechanical properties of wood. However, there are some ways to mitigate the decreasing results [14]. Thermal modification is mostly carried out between the temperatures of

160 and 260 °C. Temperatures that are lower than 140 °C usually result in very little changes in the mechanical properties, but higher temperatures result in so much degradation that the mechanical properties are usually unacceptable. Thermal modification processes that are currently used at an industrial scale are not higher than 260 °C; in practice, temperatures between 150 and 230 °C are more accepted [11,15]. The reduction of swelling in wood specimens caused by increase in temperature and duration of heat treatment was often attributed to the destruction in hemicellulose compounds [15]. However, structural modifications and chemical changes of lignin were suggested to also be involved in the process [15]. Moreover, Borrega and Kärenlampi [8] revealed that a reduction in hygroscopicity can not only be attributed to mass loss, but another mechanism that was also active in the process. They suggested that the active mechanism might be an irreversible hydrogen bonding that occurs during the process of water movement within the porous system of wood structures. This bonding was reported by other researchers to change physical and mechanical properties, as well as fluid flow in the solid woods.

High thermal conductivity coefficients of metal nanoparticles [16–20] were used in improving some of the properties in solid woods and wood-composite materials [21]. Impregnation with nanosilver suspension as well as heat treatment were also reported to alter the porous structure of solid woods, significantly altering the gas and liquid permeability [22], and possibly the penetration of paints in to the porous structure, eventually changing the adhesion strength of paints. However, authors found no or little literature on the effects of impregnation with metal nanosuspension and thermal treatment on the correlation between the paint pull-off adhesion strength with permeability in solid woods. Therefore, the present study was carried out to firstly find out the effects of nanosilver-impregnation and heat treatment on the gas permeability of beech wood, as a commercial wood species. Thereafter, and as to the nondestructive nature of permeability measurement process, pull-off strength was measured in the same specimens, providing the possibility to find out the effects of nanosilver-impregnation and heat treatment on this property, too. With regard to the fact that both of these properties, permeability and pull-off strength, depend in some way on the porous structure of the substrate, correlation between them was calculated. Moreover, as to the low thermal conductivity coefficient of wood, a separate set of specimens was impregnated with nanosilver suspension to increase thermal conductivity in the specimens and decrease the heat treatment gradient in them. This can also accelerate thermal treatment.

2. Materials and Methods

2.1. Specimen Procurement

Five discs from different beech trees (*Fagus orientalis* Lipsky) were cut at breast-height and air dried. From each disk, 80 longitudinal cylindrical specimens were prepared. The diameter of specimens was 17 mm, and the length was 30 mm. Specimens were checked not to have any knots, fissures, or cracks. They were first air dried for eight months, and then they were kept for four weeks in a conditioning room (25 ± 2 °C, and 40% ± 3% relative humidity) to avoid any undesirable effects of kiln-drying [23]. Specimens were divided into two equal groups of control (C) and nanosilver-impregnated (NS) groups. Each group was again divided into four subgroups of unheated, heat-treated at 145 °C (HT-145), heat-treated at 165 °C (HT-165), and heat-treated at 185 °C (HT-185). Gas permeability of all specimens was measured in the first phase of the research project before any heat treatment and impregnation. The NS specimens were then impregnated with a 400 ppm aqueous nanosilver suspension. All specimens were again kept in a conditioning room for two more months. Gas permeability was again measured. They were painted with three unpigmented resins of sealer-clear, polyester, and lacquer with organic solvent, produced by Pars-Eshen Co., as to their great popularity in the local market. A dolly was stuck to one end of the specimens for the paint-adhesion testing. Moisture content of specimens was 8% ± 0.5% in all treatments when permeability and pull-off tests were carried out because wood has a thermo-hygromechanical behavior and the properties relating to its deformation depend on the same factors, including moisture content, temperature, and relative humidity [24].

2.2. Pull-Off Adhesion Strength Testing

Adhesion strength testing provides the force needed to pull a test diameter of coating away from the substrate. Adhesion tests were carried out in accordance with ASTM D4541-02 [25]. In the present study, an automatic PosiTest® pull-off adhesion tester (DelFesko, NY, USA) was used (Figure 1). This was a self-aligning spherical dolly-head tester (Type V according to the ASTM standard). The diameter of the dolly used was 20 mm. The greatest tensile pull-off force for which the coating could adhere to the substrate was evaluated in terms of mega Pascal. Breaking points occurred along the weakest plane of the whole structure. The adhesion strength (*X*) was calculated in terms of MPa (Equation (1)).

$$X = \frac{4F}{\pi d^2} \quad (1)$$

where *F* is the rupture force (Newton), and *d* is the diameter of the experiment cylinder (mm) (ASTM D4541-02).

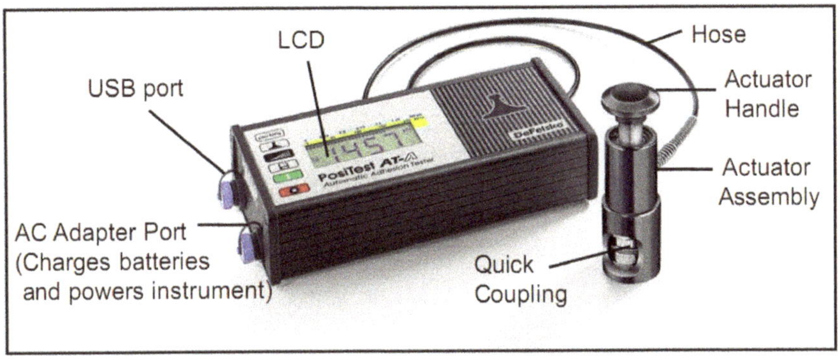

Figure 1. PosiTest® AT-A automatic pull-off adhesion tester.

The moisture content of the specimens was 8% ± 0.5% when the pull-off adhesion tests were carried out, and the temperature was 25 ± 3 °C. In order to have an estimate of the pull-off strength of the substrate material for comparison purposes, a set of specimens was also prepared without paints.

2.3. Gas Permeability Measurement

Many methods and apparatuses have been used and invented to measure permeability in solid woods and wood-composite materials as porous materials [1,26–29]. In the present study, longitudinal gas permeability was measured with an apparatus equipped with a 7-level electronic device; time measurement was carried out with millisecond precision [26,30,31]. Falling water was applied to measure and calculate the specific longitudinal gas permeability values. For each single specimen, gas permeability values were separately measured at seven different vacuum pressures. In every single run, the seven time measurements were carried out to finally calculate seven specific permeability values for each specimen. The glass tube had an internal diameter of 13 mm. Water level was put to more than 15 cm above the first time measurement section (Gas 1). A fully airtight connection was made between the specimen and holder. The pressure difference (ΔP) was monitored by a pressure gauge connected to the permeability apparatus. The pressure difference could be read at any particular time and height. This provided monitoring of the viscose flow [22]; the gauge had a precision of millibar. Vacuum pressures at the starting and stopping levels were also measured [30].

Each specimen was tested three times to finally calculate the permeability. Then, the superficial permeability coefficient was calculated (Equations (2) and (3)) [32,33]. The superficial gas permeability coefficients (k_g) were corrected by the viscosity of air ($\mu = 1.81 \times 10^{-5}$ Pa s) for the calculation of the specific gas permeability ($K_g = k_g\mu$).

$$K_g = \frac{V_d CL(P_{atm} - 0.074\bar{z})}{tA(0.074\bar{z})(P_{atm} - 0.037\bar{z})} \times \frac{0.760\text{mHg}}{1.013 \times 10^6\text{Pa}} \tag{2}$$

$$C = 1 + \frac{V_r(0.074\Delta z)}{V_d(P_{atm} - 0.074\bar{z})} \tag{3}$$

where:

k_g = superficial gas permeability ($m^3\ m^{-1}$);
$V_d = \pi r^2 \Delta z$ [r = radius of measuring tube (m)] (m^3);
C = correction factor for gas expansion as a result of change in static head and viscosity of water;
L = length of wood specimen (m);
P_{atm} = atmospheric pressure (mHg);
$\bar{z}$ = average height of water over surface of reservoir during period of measurement (m);
t = time (s);
A = cross-sectional area of wood specimen (m^2);
Δz = change in height of water during time t (m);
V_r = total volume of apparatus above point 1 (the volume of hoses was included) (m^3).

2.4. Nanosilver Impregnation

A 400 ppm aqueous nanosilver suspension was produced via the electrochemical technique in cooperation with Mehrabadi Mfg. Co. (Tehran, Iran). The nanoparticles ranged 30–80 nm in size; the pH was 6–7. Anionic and cationic surfactants were used to stabilize nanoparticles in the suspension. Specimens were impregnated using an empty-cell process in a pressure vessel. The pressure was set at 2.5 bars for 25 min. Before conducting the tests on the specimens, they were kept for three months at room temperature.

The nanosuspension was prepared by transferring the silver metal ion from the aqueous phase to the organic phase, where it reacted with a monomer. The formation and size of the nanosilver was carefully monitored by transmission electron microscopy (TEM). Samples for TEM were prepared by drop-coating the Ag nanoparticle suspensions on to carbon-coated copper grids. Micrographs were obtained using an EM-900 ZEISS transmission electron microscope (Carl Zeiss AG, Jena, Germany). Two kinds of surfactants (anionic and cationic) were used in the suspension as stabilizer. The concentration of the surfactants was two times the nanosilver [34].

2.5. Heat Treatment Process

Specimens to be heat-treated were randomly placed in an oven. Heat treatment was carried out at atmospheric pressure and with normal air. The starting moisture content of specimens was 8% ± 0.5%. Heat treatment of each thermal modification temperature level was carried out in a single run. All HT and NS-HT specimens were heat-treated at 145 °C for 12 h. Heat treatment for the HT-145 and NS-HT-145 specimens was discontinued. Then, HT-165 and NS-HT-165, as well as HT-185 and NS-HT-185 specimens continued to be heated at 165 and 185 °C for four more hours, respectively. During the heat treatment process, no specimen was in touch with the metal parts of the oven, to prevent extra overheating at one spot. Once the heat treatment process was completed, the silicone adhesive around all specimens was checked to make sure there was no failure in them. The gas permeability was then measured.

2.6. SEM Imaging

Scanning electron microscope (SEM) imaging was carried out at a thin-film laboratory, FE-SEM lab (Field Emission), School of Electrical and Computer Engineering, the University of Tehran. A field-emission cathode in the electron gun of a scanning electron microscope provides narrower probing beams at low as well as high electron energies. This way, the spatial resolution was improved and the charging and damage to the specimens were minimized. As wood is a nonconductive material, a gold sputtering thickness of 6–8 nm was applied on the surface of specimens prior to SEM imaging.

2.7. Statistical Analysis

One-way analysis of variance (ANOVA) was carried out to discern significant difference at 95% level of confidence, using the SAS software program (version 9.2) (2010). Grouping was then made between treatments using Duncan's multiple range test. Regression analyses, including dendrogram and using Ward methods with squared Euclidean distance intervals, were carried out by SPSS/18 (2010). For the regression analysis, a *p*-value of less than 5% (<0.05) was determined as the significance level. Based on the *p*-value, the critical *R* value was 0.63. Fitted-line and scatter plots were made using Minitab software (version 16.2.2) (2010).

3. Results

The amount of nanosilver suspension absorption after the impregnation process was measured to be 0.38 g/cm^3. Results of the permeability tests showed that the lowest specific gas permeability was observed in specimens before NS-impregnation or heat treatment (NS-HT-165 treatment, 6.898×10^{-13} m^3 m^{-1}), and the highest was found in NS-HT-185 after NS-impregnation (15.576×10^{-13} m^3 m^{-1}).

In the un-impregnated specimens, heat treatment slightly increased the specific gas permeability values in all three temperatures of 145, 165, and 185 °C, but the increase was not statistically significant (Figure 2). The highest increase in the un-impregnated specimens was observed in HT-145 treatment (only 5.3%). As for the nanosilver-impregnated specimens (the right columns of Figure 2), impregnation with the nanosuspension significantly increased the permeability in all treatment (Figure 2). In some cases, the amount of increase was more than 107.5%. The highest specific gas permeability was found in the NS-HT-185 treatment after the NS-impregnation and before the heat treatment. The increasing effect of heat treatment on permeability was significantly intensified by the thermal conductivity of silver nanoparticles; the highest increase caused by heat treatment of NS-impregnated specimens was seen in NS-HT-145 treatment (20.8%). The only treatment showing a decreasing trend in permeability caused by heat treatment was NS-HT-185. This treatment showed a decrease of about 8.8%.

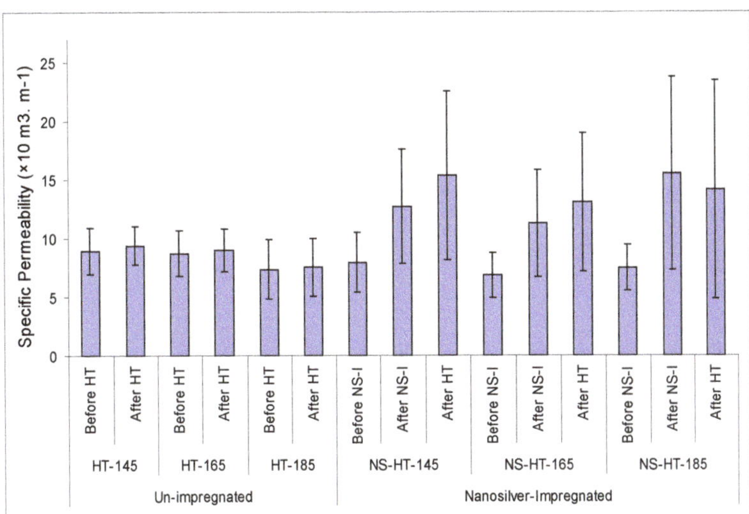

Figure 2. Specific longitudinal gas permeability in the beech specimens ($\times 10^{-13}$ m^3 m^{-1}) (error bars indicate the standard deviation for each column).

The maximum pull-off adhesion strength was observed in the control specimens (un-impregnated and un-heated) painted with polyester (8.98 MPa), and the lowest pull-off strength was found in the un-painted and un-heated specimens that were NS-impregnated (3.10 MPa) (Figure 3). NS-impregnation generally increased pull-off strength in the heat-treated painted specimens. In the unpainted specimens, however, a reverse trend was observed.

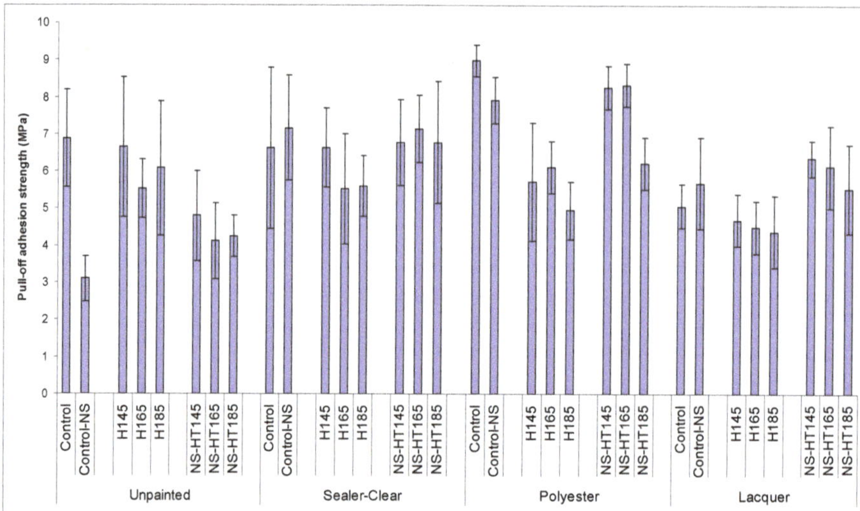

Figure 3. Pull-off adhesion strength (MPa) of different treatments of beech specimens (error bars indicate the standard deviation for each column).

4. Discussion

Heat treatment made the specimens lose their moisture content completely, resulting in the vessel perforations and pit openings to shrink and be wide open. This shrinkage led to increased specific air permeability, because permeability is influenced by the porous structure of the material, and even slight changes in the porous system may significantly affect permeability [35,36]. The shrinkage is partially permanent due to the irreversible hydrogen bonding that occurs when water is moved in the cell wall [8,21].

NS-impregnation also resulted in significant increases of permeability in all NS-impregnated treatments. This increase was attributed to the high pressure applied in the pressure vessel, resulting in the rupture of vessel perforations, tyloses, and any other physical obstacle in the way of fluid transfer. SEM images showed vessels that were blocked (Figure 4A). The blockage was ruptured and torn open after NS-impregnation (Figure 4B,C). Therefore, fluid could flow more easily, and eventually, permeability increased significantly.

Heat treatment decreased pull-off strength in all three coatings studied here (Figure 3). It is reported that heat treatment results in the occurrence of microcracks in the wood structure and thermal degradation of cell wall polymers [5,37–39]. These microcracks led to unwanted penetration of adhesive film into wood texture, far from being involved in the process of anchoring and sticking dolly to wood substrate. However, in the NS-impregnated specimens, heat was transferred to deeper parts of specimens, and therefore, accumulation of heat did not occur on the surface of specimens, eventually decreasing microcracks in NS-impregnated specimens. This was translated into higher pull-off strength in NS-impregnated specimens of all three types of coatings.

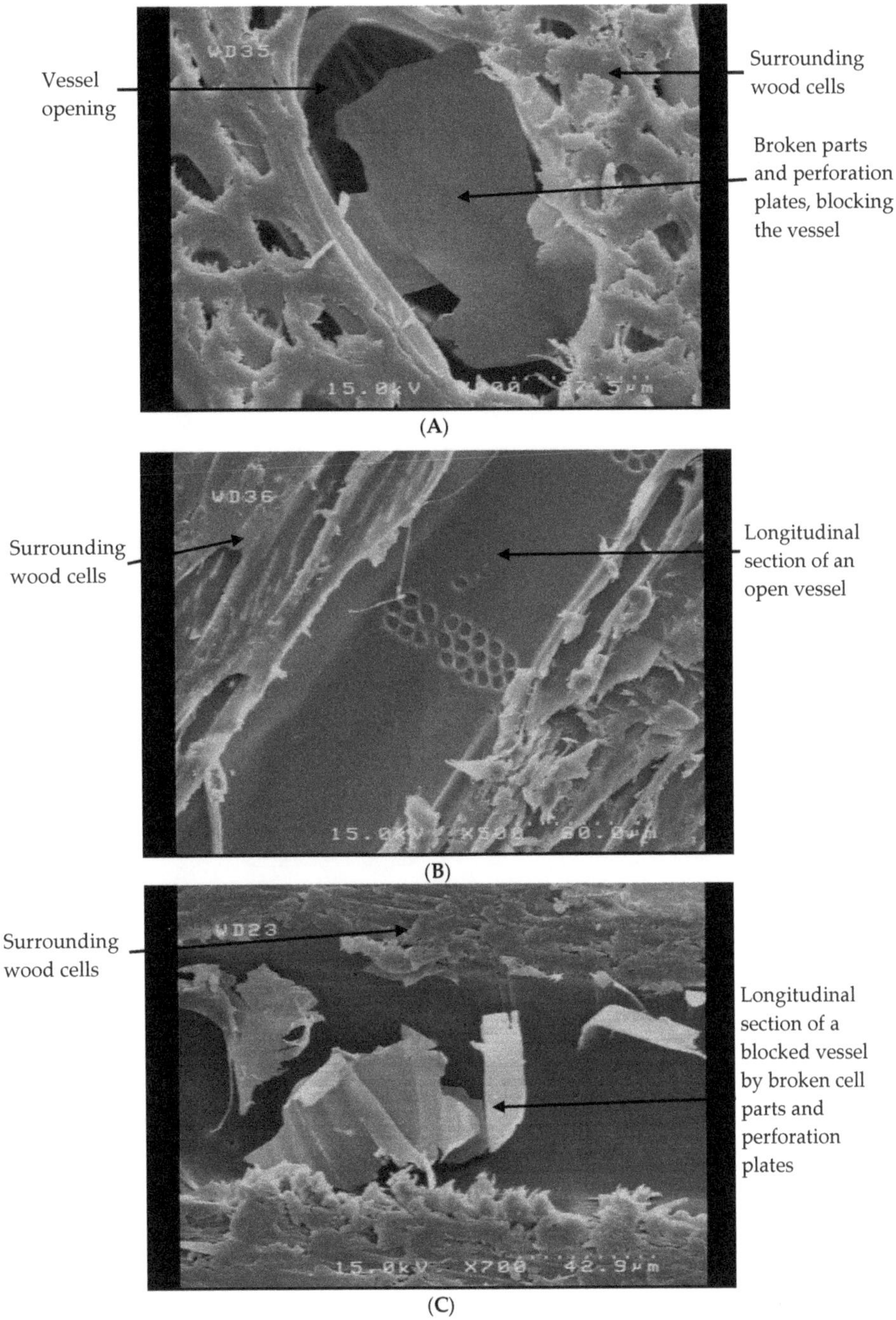

Figure 4. SEM images showing cell parts and perforation plates (↓); (**A**) cross-section of a blocked vessel by distorted and ruptured vessel elements; (**B**) longitudinal section of an open vessel; (**C**) longitudinal section of a vessel blocked by broken cell parts and perforation plates.

The fitted-line plot between pull-off adhesion strength versus specific air permeability in all treatments showed no particular trend between these two properties (Figure 5). Regression analysis also showed insignificant R^2 between air permeability versus pull-off strength in almost all specimens and treatments (Table 1). The only significant R^2 was found in HT-145 polyester specimens. However, this one case cannot be a reliable indicator as to the existence of potentially significant R^2 between the two properties of permeability and pull-off strength. Therefore, it can be concluded that air permeability cannot be considered a good criterion to estimate the pull-off strength in beech wood. Both properties (permeability and pull-off strength) are dependent on the porous structure of materials, but permeability is influenced by the continuous pores while pull-off strength is more dependent on the surface pores, whether continuous or isolated. That is, an isolated and blocked vessel can also be active in the penetration and anchoring of adhesive, similar to a continuous vessel (Figure 4A). However, permeability is only dependent on the number of continuous vessels and pores, their attributes, and the way they are connected to one another.

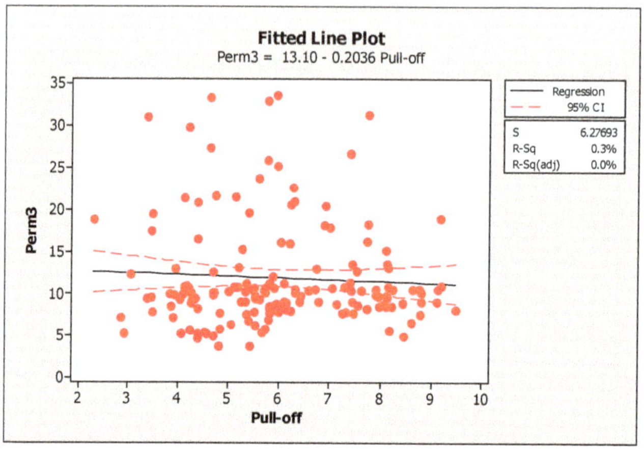

Figure 5. Fitted-line plot between pull-off adhesion strength versus specific air permeability (Perm3 = specific air permeability measured at the third vacuum pressure or water column level).

Table 1. Regression analysis between longitudinal specific air permeability versus pull-off adhesion strength in different treatments.

Treatments		*R*-Square			
		Un-Painted	SC	Lacquer	Polyester
Un-heated	Control	0.795 ns(-)	0.191 ns(+)	0.047 ns(+)	0.732 ns(+)
	NS-I	0.304 ns(-)	0.519 ns(+)	0.443 ns(+)	0.200 ns(+)
HT-145	Control	0.004 ns(-)	0.683 ns(+)	0.503 ns(-)	0.954 *(-)
	NS-I	0.004 ns(-)	0.061 ns(+)	0.034 ns(-)	0.186 ns(-)
HT-165	Control	0.111 ns(-)	0.470 ns(+)	0.105 ns(+)	0.676 ns(+)
	NS-I	0.469 ns(-)	0.135 ns(+)	0.650 ns(-)	0.284 ns(-)
HT-185	Control	0.240 ns(+)	0.638 ns(+)	0.234 ns(+)	0.438 ns(+)
	NS-I	0.551 ns(+)	0.520 ns(-)	0.246 ns(-)	0.481 ns(+)

SC = sealer-clear painted; NS-I = nanosilver-impregnated; ns = nonsignificant; HT = heat-treated.

5. Conclusions

- Impregnating beech wood with nanosilver suspension results in the rupture of perforation plated and pit openings, and eventually permeability increases significantly.
- Higher permeability causes more adhesive being penetrated deep in to the pores of the wood substrate, being left out of the sticking process, eventually decreasing pull-off strength.
- Permeability depends on the continuous pore system. Pull-off adhesion strength, however, is dependent on the surface pore system of the wood substrate. Therefore, there is no significant correlation between pull-off adhesion strength versus air permeability.

Author Contributions: Methodology, H.T. and A.E.; Validation, H.T.; Investigation, H.T. and A.E; Writing-Original Draft Preparation, H.T., A.E. and A.P.; Writing-Review & Editing, H.T., A.E. and A.P.; Visualization, H.T. and A.E; Supervision, H.T and A.P.

Funding: This research received no external funding.

Acknowledgments: Authors are grateful to Jack Norton (Retired, Horticulture & Forestry Science, Queensland Department of Agriculture, Forestry and Fisheries, Australia) for his constant scientific support and inspiration, as well as Alexander von Humboldt Stiftung, Germany.

Conflicts of Interest: The authors declare no conflict of interest.

References

1. Siau, J.F. *Flow in Wood*; Syracuse University Press: Syracuse, NY, USA, 1971.
2. Siau, J.F. *Wood: Influence of Moisture on Physical Properties*; Blacksburg, V.A., Ed.; Department of Wood Science and Forest Products, Virginian Polytechnic Institute and State University: Blacksburg, VA, USA, 1995.
3. Siau, J.F. *Transport Processes in Wood*; Springer: Berlin/Heidelbergh, Germany; GmbH & Co. KG: Berlin, Germany, 2011.
4. Skaar, C. *Wood-Water Relations*; Springer: Berlin/Heidelberg, Germany, 1988.
5. Oltean, L.; Teischinger, A.; Hansmann, C. Influence of temperature on cracking and mechanical properties of wood during wood drying—A review. *BioResources* **2008**, *2*, 789–811.
6. Mosqueda, M.R.P.; Tabil, L.G.; Christensen, C. Effect of drying conditions and level of condensed distillers solubles on protein quality of what distillers dried grain with solubles. *Dry. Technol.* **2013**, *31*, 811–824. [CrossRef]
7. Fernandez-Puratich, H.; Oliver-Villanueva, J.V. Quantification of biomass and energetic value of young natural regenerated stands of *Quercus ilex* under Mediterranean conditions. *Bosque* **2014**, *35*, 65–74.
8. Borrega, M.; Karenlampi, P.P. Hygroscopicity of heat-treated norway spruce (*Picea abies*) wood. *Eur. J. Wood Wood Prod.* **2010**, *68*, 233–235. [CrossRef]
9. Hering, S.; Keunecke, D.; Niemz, P. Moisture-dependent orthotropic elasticity of beech wood. *Wood Sci. Technol.* **2012**, *46*, 927–938. [CrossRef]
10. Barna, M. Natural regeneration of *Fagus sylvatica* L.: A Review. *Austrian J. For. Sci.* **2011**, *128*, 71–91.
11. Hill, C. *Wood Modification Chemical, Thermal and Other Processes*; John Wiley & Sons, Ltd.: Hoboken, NJ, USA, 2006.
12. Schmidt, O. *Wood and Tree Fungi: Biology, Damage, Protection, and Use*; Springer: Berlin/Heidelberg, Germany, 2006.
13. Maresi, G.; Oliveira Longa, C.M.; Turchetti, T. Brown rot on nuts of *Castanea sativa* Mill: An emerging disease and its causal agent. *iForest Biogeosci. For.* **2013**, *6*, 294–301. [CrossRef]
14. Awoyemi, L. Determination of optimum borate concentration for alleviating strength loss during heat treatment of wood. *Wood Sci. Technol.* **2007**, *42*, 39–45. [CrossRef]
15. Repellin, V.; Guyonnet, R. Evaluation of heat treated wood swelling by differential scanning calorimetry in relation with chemical composition. *Holzforschung* **2005**, *59*, 28–34. [CrossRef]
16. Pati, R. Molecule for electronics: A myriad of opportunities comes with daunting challenges. *J. Nanomater. Mol. Nanotechnol.* **2012**, *1*. [CrossRef]
17. Sadeghi, B.; Rastgo, S. Study of the shape controlling silver nanoplates by reduction process. *Int. J. Bio-Inorg. Hybrid Nanomater.* **2012**, *1*, 33–36.

18. Saber, R.; Shakoori, Z.; Sarkar, S.; Tavoosidana, G.; Kharrazi, S.; Gill, P. Spectroscopic and microscopic analyses of rod-shaped gold nanoparticles interacting with single-stranded DNA oligonucleotides. *IET Nanobiotechnol.* **2013**, *7*, 42–49. [CrossRef] [PubMed]
19. Tomic, S.L.; Micic, M.M.; Dokic, D.; Vasiljevic-Radvic, D.G.; Filipovic, J.M.; Suljovrujic, E.H. Preparation of silver (I) complexes with itaconic acid-based hydrogels for biomedical application. *Mater. Manuf. Process.* **2009**, *24*, 1197. [CrossRef]
20. Nguyen, N.T.; Nguyen, B.H.; Ba, D.T.; Pham, D.G.; Khai, T.V.; Nguyen, L.T.; Tran, L.D. Microwave-assisted synthesis of silver nanoparticles using chitosan: A novel approach. *Mater. Manuf. Process.* **2014**, *29*, 418. [CrossRef]
21. Taghiyari, H.R.; Enayati, A.; Gholamiyan, H. Effects of nano-silver impregnation on brittleness, physical and mechanical properties of heat-treated hardwoods. *Wood Sci. Technol.* **2013**, *47*, 467–480. [CrossRef]
22. Taghiyari, H.R. Effects of heat-treatment on permeability of untreated and nanosilver-impregnated native hardwoods. *Maderas Cienc. Y Tecnol.* **2013**, *15*, 183–194. [CrossRef]
23. Sönmez, A.; Budakci, M.; Huseyin, P. The effect of the moisture content of wood on the layer performance of water-borne varnishes. *BioResources* **2011**, *6*, 3166–3178.
24. Figueroa, M.; Bustos, C.; Dechent, P.; Reyes, L.; Cloutier, A.; Giuliano, M. Analysis of rheological and thermo-hygro-mechanical behaviour of stress-laminated timber bridge deck in variable environmental conditions. *Maderas Cienc. Y Tecnol.* **2012**, *14*, 303–319.
25. *ASTM D 4541-02. Standard Test Method for Pull-Off Strength of Coatings Using Portable Adhesion Testers*; ASTM: West Conshohocken, PA, USA, 2006.
26. Taghiyari, H.R.; Kalantari, A.; Avramidis, S. Effect of wollastonite nanofibers and exposure to *Aspergillus niger* fungus on air flow rate in paper. *Measurement* **2019**, *136*, 307–313. [CrossRef]
27. Shi, S.H.Q. Diffusion model based on Fick's second law for the moisture absorption process in wood fiber-based composites: Is it suitable or not? *Wood Sci. Technol.* **2007**, *41*, 645–658. [CrossRef]
28. Dermoe, D.; Zillig, W.; Carmeliet, J. Variation of measured cross-sectional cell dimensions and calculated water vapor permeability across a single growth ring of spruce wood. *Wood Sci. Technol.* **2012**, *46*, 827–840. [CrossRef]
29. Shibata, H.; Hirohashi, Y. Effect of segment scale in a pore network of porous materials on drying periods. *Dry. Technol.* **2013**, *31*, 743–751. [CrossRef]
30. Ghorbani, M.; Akhtari, M.; Taghiyari, H.R.; Kalantari, A. Effects of silver and zinc-oxide nanoparticles on gas and liquid permeability of heat-treated Paulownia wood. *Austrian J. For. Sci.* **2012**, *129*, 106–123.
31. Esmailpour, A.; Taghiyari, H.R.; Golchin, M.; Avramidis, S. On the fluid permeability of heat treated paulownia wood. *Int. Wood Prod. J.* **2019**, *10*, 55–63. [CrossRef]
32. Hernandez, V.; Avramidis, S.; Navarrete, J. Albino strains of *Ophiostoma* spp. Fungi effect on radiate pine permeability. *Eur. J. Wood Wood Prod.* **2012**, *70*, 551–556. [CrossRef]
33. Choo, A.C.Y.; Tahir, M.P.; Karimi, A.; Bakar, E.S.; Abdan, K.; Ibrahim, A.; Balkis, F.A.B. Study on the longitudinal permeability of oil palm wood. *Ind. Eng. Chem. Res.* **2013**, *52*, 9405–9410. [CrossRef]
34. Taghiyari, H.R.; Norton, J. Effect of silver nanoparticles on hardness in medium-density fiberboard (MDF). *J. iForest Biogeosci. For.* **2014**, *8*, 677–680. [CrossRef]
35. Dashti, H.; Shahverdi, M.; Taghiyari, H.R.; Salehpur, S.; Heshmati, S. Effects of steaming and microwave pretreatments on mass transfer characteristics of Aleppe oak (*Quercus infectoria*). *BioResources* **2012**, *7*, 3262–3273.
36. Taghiyari, H.R. Correlation between gas and liquid permeability in some nanosilver-impregnated and untreated hardwood. *J. Trop. For. Sci.* **2012**, *24*, 249–255.
37. Korkut, S.; Budakci, M. Effect of high-temperature treatment on the mechanical properties of Rowan (*Sorbus aucuparia* L.) wood. *Dry. Technol.* **2009**, *27*, 1240–1247. [CrossRef]
38. Bayani, S.; Taghiyari, H.R.; Papadopoulos, A.N. Physical and mechanical properties of thermally-modified beech wood impregnated with silver nano-suspension and their relationship with the crystallinity of cellulose. *Polymers* **2019**, *11*, 1538. [CrossRef] [PubMed]
39. Bayani, S.; Bazyar, B.; Mirshokraie, S.A.; Taghiyari, H.R. Effects of heat treatment on the relative amounts of cellulose in nanosilver-impregnated and untreated poplar wood (*Populus alba*). *Floresta E Ambiente* **2019**, *26*, e20160398. [CrossRef]

coatings

MDPI

Article

Comparison of Exterior Coatings Applied to Oak Wood as a Function of Natural and Artificial Weathering Exposure

Eliška Oberhofnerová [1], Kristýna Šimůnková [1], Ondřej Dvořák [1], Irena Štěrbová [1], Salim Hiziroglu [2], Přemysl Šedivka [1] and Miloš Pánek [1,*]

[1] Department of Wood Processing and Biomaterials, Faculty of Forestry and Wood Sciences, Czech University of Life Sciences in Prague, Kamýcká 129, 165 00 Prague, Czech Republic; Eliska.Oberhofnerova@seznam.cz (E.O.); simunkovak@fld.czu.cz (K.Š.); dvorak18@fld.czu.cz (O.D.); sterbovai@fld.czu.cz (I.Š.); sedivka@fld.czu.cz (P.Š.)

[2] Department of Natural Resource Ecology and Management, Oklahoma State University, 303-G Agricultural Hall, Stillwater, OK 74078, USA; salim.hiziroglu@okstate.edu

* Correspondence: panekmilos@fld.czu.cz; Tel.: +420-224-383-867

Received: 13 November 2019; Accepted: 14 December 2019; Published: 16 December 2019

Abstract: Artificial weathering can significantly reduce the testing time needed for proving coating durability, nevertheless its reliability is still not thoroughly proven. In this study, eight different transparent and pigmented coating systems, namely oil, acrylate, alkyd and urethane alkyd were evaluated through natural and artificial weathering tests on oak samples by measuring colour, gloss and surface wettability and by macroscopic and microscopic evaluation. The oil coatings performed well in wood colour stability evaluations, while the best gloss and wettability change results were noted for acrylate coatings. Pigmented coatings were characterized by significantly lower colour changes than transparent ones. The gloss and wettability changes were more sensitive to coating disruption than to total colour changes of coated wood associated with chemical changes in wood. The findings in this work showed that values of gloss changes and surface wettability for all types of coatings exposed to artificial and natural weathering resulted in significant differences from each other. The data obtained by artificial weathering method provide basic results of coatings durability and, ideally, natural weathering should be performed at the same time to support the results from laboratory tests by exposing wood under real conditions.

Keywords: artificial weathering; coatings; durability; natural weathering; oak wood

1. Introduction

Oak (*Quercus petraea* L.) wood is often used for the exterior applications, mostly in construction of bridges, pergolas, balconies or garden furniture, where higher natural durability [1] is required. Oak contains a relatively high amount of phenol extractives, mainly vescalagin, castalagin, gallic and ellagic acids [2], creating problems in the field of surface treatment durability [3,4]. Tannins in oak wood also retard coating hardening [5]. The complex open vessel morphological structure of oak wood complicates the overall application of coatings. A photodegradation process of oak wood accompanied with significant discolouration and leaching of extractives from the surface takes place during the initial phases of outdoor exposure [6], more intensely in the heartwood zone [7], which leads to the need to protect oak wood by coatings to maintain its natural appearance.

Exterior wood coatings are used to improve the properties of substrate wood [8], reduce the effects of degradation factors [9–11] and prolong the service life of the material. The exterior coating generally protects against moisture uptake and related dimensional changes, protects against photochemical degradation, and prevents microbiological degradation [12,13]. The problems with transparent coatings have been well discussed by Evans et al. [13]. Their advantage is the ability to protect wood and preserve the natural look and colour [13], but they have the disadvantage of not protecting the substrate wood against UV and visible light radiation as well as pigmented coatings [14]. The different performance of coatings is also caused by their polymer base [15], the type of solvent [16] or even by the underlying wood species [17,18]. Coatings protect underlying wood, but they are exposed to weathering process causing their degradation [17,19,20]. The durability of coatings against atmospheric degradation is assessed via natural weathering (NW) or artificial weathering (AW) tests [8,21–25] with parameters given in international standards. The older and more common method is natural weathering [26], which provides reliable results of coating durability due to the synergistic action of outdoor factors. Accelerated artificial weathering [27] is carried out in laboratory conditions, to simulate the exterior environment [22,28]. Artificial weathering can significantly reduce the required testing time, however its reliability is still often questioned. Correlations between weathering methods were done in several studies [23,29–31] however the results were ambiguous. Valverde and Moya [32] developed a model to predict total colour difference between natural weathering and accelerated weathering for different kinds of finishes of three tropical species. Cogulet et al. [11] focused on how the impacts of different weathering methods challenges the reliability of AW and states that it is necessary to test coating systems in an end-use environment for accurate assessment of their likely performance.

In these works, where wood weathering was studied, more qualitative parameters of wood coatings during both exposures were evaluated–change of colour parameters [3,11,30], coating thickness [23,30], coating adhesion [30,33] or gloss [3,34]. Especially the change of colour during weathering serves as a basic indicator of the rate of degradation [18,22,30]. In a study of Moya et al. [30], colour change was higher in all species after NW than after AW due to the constant variation of solar radiation, moisture, water, air contamination and biotic agents, which accelerates colour degradation processes [8,35]. These findings are consistent with previous studies indicating the difficulty of reproducing the synergistic action of NW factors during AW [11,17,23,24,36].

Currently there is very limited information on characteristics of transparent and pigmented coatings on oak wood when they are exposed to both natural and artificial weathering. Therefore, the objective of this study was to compare the performance of eight different transparent and pigmented coating systems applied on oak samples using natural and artificial weathering tests. The efficiency of specific coating systems was determined by measurements of colour, gloss and surface wettability changes and by regular visual macroscopic and microscopic evaluation of the samples.

2. Materials and Methods

2.1. Sample Preparation, Coatings and Weathering Process

Samples of oak (*Quercus petraea* L.) wood harvested in the Czech Republic having an average oven dry density of ρ_0 = 705 kg/m^3 [37] were used for the experiment. The samples were conditioned to 12 ± 2% moisture content. Test samples were prepared from the heartwood zone and they were visually sorted in order to minimalize the colour variability of the tested wood materials. The dimensions of the samples were 375 mm × 78 mm × 20 mm for natural weathering and 45 mm × 45 mm × 8 mm for the artificial weathering tests. Tangential surfaces were exposed to weathering in both cases. Two samples for NW and four samples for AW for each type of treatment were used.

Eight different transparent or pigmented coatings were applied to the samples based on the producer's recommendation. Their specification and application details are given in Table 1. One group of samples was left untreated as control samples to compare coatings performance on treated samples. The cross ends of samples were sealed with silicon to minimize additional water uptake.

Table 1. Specifications of the tested coatings according to the producers.

Coating Symbol	Coating Specification	Type of Coating	Transparent (T) or Pigmented (P)	Number of Layers (Dry Film Thickness)
CS	Control reference (REF) native samples without any coating system	-	-	-
AC1	Acrylate thick-layer water-based stain with fungicides (5-chloro-2-methylisothiazol-3(2*H*)-one)	Acrylate	T	2 (40 μm)
AC2	Acrylate thin-layer water-based coating with a UV light absorber (1,2-benzoisothiazol-3(2*H*)-one), IPBC as fungicide	Acrylate	T	3 (20 μm)
AL1	Thixotropic alkyd coating with microparticles as a UV-stabilizer	Alkyd	T	3 (40 μm)
AL2	Thick-layer mixture of alkyds and oils with IPBC and pigments	Alkyd/Oil	P	2 (40 μm)
O1	Thin-layer oil-based with micronized pigments (TiO_2) and fungicides (propiconazole < 1%)	Oil	P	2 (10 μm)
O2	Thin-layer oil-based with dark micronized pigments (Fe_2O_3) and fungicides (propiconazole < 1%)	Oil	P	2 (5 μm)
O3	Oil-based coating with fungicides (propiconazole 0.5%)	Oil	T	3 (10 μm)
S1	Thin-layer solvent-based stain—urethane alkyds with additives in white spirit with IPBC	Urethane alkyd	P	2 (25 μm)

Note: IPBC = 3-iodo-2-propynyl butylcarbamate.

The natural weathering (NW) test was performed at Suchdol, Prague (50°07′49.68″ N, 14°22′13.87″ E) for 12 months. The climatic conditions during exposure are given in Table 2. The samples were exposed at a 45° inclination, facing south, and placed approximately 1 m above the ground according to the procedure previously described [26].

Table 2. Climatic conditions during NW. use a period (.) for decimals.

	Period of Outdoor Exposure in 2018 (months)											
Measured data per day	**1**	**2**	**3**	**4**	**5**	**6**	**7**	**8**	**9**	**10**	**11**	**12**
Average temperature (°C)	3.5	−1.9	2.1	13.8	17.4	18.7	21.6	22.0	16.0	10.7	4.7	2.9
Average relative humidity (%)	81.2	73.4	70.3	57.8	59.7	63.4	49.2	53.4	64.0	70.3	84.5	82.2
Total precipitation (mm)	17.2	4.6	29.1	14.2	20.8	87.9	8.8	56.5	43.5	23.9	6.5	40.6
Average solar radiation (kJ/m^2)	2432	6473	8305	17,365	21,428	20,253	22,177	18,250	12,455	7915	3280	1992

Note: based on the data from http://meteostanice.agrobiologie.cz [38].

The artificial weathering (AW) test was performed in UV-chamber QUV (Q-Lab, Cleveland, OH, USA) according to a modified EN 927-6 method [27]. The total time consisted of six cycles (1008 h) of weathering in the UV chamber and 36 h of temperature cycling. During the each weekly cycle of irradiation and spraying, the samples were transferred to a Discovery My DM340 conditioning chamber (ACS, Massa Martana, Italy) and exposed to three cycles each lasting 2 h of temperature changes from −25 to +80 °C (with 25% RH). The alternation of UV radiation, spray, and low temperature cycles, leading to a better imitation of the exterior conditions in Central and Northern Europe, was previously used by Van den Bulcke et al. and Pánek et al. [22,39].

2.2. Colour Change (ΔE*) Test

Colour variations of the specimens were evaluated through natural and artificial weathering exposure of oak samples with 8 different coatings. The colour parameters L*a*b* [40] of the test specimens were measured after 0, 6 and 12 months of NW and after 0, 1, 3 and 6 weeks of AW using CM-600d Spectrophotometer (Konica Minolta, Osaka, Japan). For the observation of reflection, the specular component was included (SCI mode) at a 10° angle and d/8 geometry with an illumination standard of D65 (corresponding to daylight in 6500 K). Six measurements of each tested sample were carried out for each weathering time. Colour changes evaluations were done in CIE L*a*b* colour space where L* is lightness from 0 (black) to 100 (white); a* is chromaticity coordinate + (red) or – (green); b* is chromaticity coordinate + (yellow) or – (blue).

The total colour difference ΔE* [40] was subsequently calculated from relative changes of colour (ΔL*, Δa*, and Δb*) using Equation (1):

$$\Delta E^* = \sqrt{(\Delta L^*)^2 + (\Delta a^*)^2 + (\Delta b^*)^2} \quad (1)$$

2.3. Gloss Change (ΔG*) Test

The gloss of the different coatings before and during weathering tests was measured using MG268-F2 glossmeter (KSJ, Quanzhou, China) on the basis of [41]. Six measurements at a 60° angle per sample after 0, 6 and 12 months of NW and after 0, 1, 3, and 6 weeks of AW were carried out to evaluate gloss changes of the samples.

2.4. Surface Wettability Change (ΔW*) Test

The sessile drop method with static contact angle measurement was performed using a Krüss DSA 30E goniometer (Krüss, Hamburg, Germany) with the methodology used in previous studies [42,43]. Twenty measurements were taken for each sample, with distilled water drops with a dosing volume of 5 μL. The contact angle values were determined after 5 s of drop deposition on surface of the sample before weathering and after 0, 6 and 12 months of NW and after 0, 1, 3 and 6 weeks of AW.

2.5. Macroscopic and Microscopic Evaluation

Tested surfaces of the samples were regularly macroscopically evaluated using a Canon 2520 MFP scanner with 300 DPI resolution (Canon, Tokyo, Japan). Creations of cracks, defoliation of coating systems were visually analysed. Microscopic structural changes of coatings and surface of the samples, creation of ruptures, 3D-images of surface profiles were also studied employing confocal laser scanning microscope Lext Ols 4100 (Olympus, Tokyo, Japan) with 108-fold magnification.

2.6. Statistical Analysis

Data were analysed using MS Excel (Microsoft, Redmond, WA, USA) and STATISTICA 13.2 (StatSoft, Palo Alto, CA, USA) using mean values, bar graphs and ANOVA for analysing the statistical significance of selected factors with significance level $\alpha = 0.05$. Spearman rank correlation between NW after 12 months and AW after 6 weeks on the basis of ΔE*, ΔL*, ΔG* and ΔW* values of tested coatings was also analysed. The Spearman rank correlation coefficient was calculated by Equation (2):

$$Rho_S = 1 - \left[\frac{6 \cdot \sum (\text{Rank Difference})^2}{n^3 - n} \right] \quad (2)$$

where n is number of items evaluated.

3. Results and Discussion

Results on coated samples exposed to artificial and natural weathering showed different behaviours. The type of coating system applied on oak wood samples has a statistically significant effect ($p < 0.05$) on the evaluated properties both during AW and NW (Table 3).

Table 3. Statistical evaluation of significance of selected factors.

		ΔL *	ΔE *	ΔG *	ΔW *
Type of coating system	AW	0.000 *	0.001 *	0.000 *	0.000 *
	NW	0.000 *	0.000 *	0.000 *	0.000 *

* signifies $p < 0.05$ (statistically significant at significance level 0.05).

3.1. Colour Change of the Samples

Total colour difference ΔE* calculated from measured colour parameters was the main indicator representing coating durability during weathering [18,22,30]. The specific colour parameters (L*, a*, b*) describe the colour change more closely. During AW, the observed increase of values a* and decrease of b* values showed a tendency of the wood surface to turn reddish and become less yellow shade. Decreases of both a* and b* parameters were observed during NW, which indicates the opposite trend. But in most cases, ΔE* was affected mainly by changes in lightness (ΔL*) as in the study of Oberhofnerová et al. [31]. Changes in lightness of different coating during weathering is illustrated in Figure 1. There are obvious differences in lightness parameters based on the weathering type-decrease (negative value) of lightness during AW (except reference samples) indicating a tendency to turn into darker and increase during NW indicating lightening. This trend is caused by the different weathering conditions and ratio of degradation and leaching of photodegraded extractives and lignins observed mainly on the transparent tested coatings. NW differing also in the presence of mould and dust and other pollution in exterior which infiltrate in the degraded surface of wood or coating [10]. These conditions are not simulated in laboratory testing [21,43]. But darkening of the surfaces caused by the action of pollutants and moulds was negligible during this period of NW and mainly leaching of darker oak extractives and changes in pigmented coatings caused increasing of L* parameter.

Total colour difference (ΔE*) of tested coatings, was closely linked to lightness changes. Control samples manifested the highest colour differences during both types of weathering. Total colour difference values are characterized by a systematic increase during exposure [43,44], with higher changes during initial phases of weathering [28,43]. In this study, only coatings AC2 and AL1 followed this trend during AW (Figure 2). The lowest colour difference was noticed for O1 during AW and O2 during NW, which is in accordance with lightness changes in Figure 1. Those were the only coatings able to protect the wood to the extent of ΔE* < 3, which is considered as a low colour difference that cannot be distinguished by a subjective observer [45]. These oil coatings differed from each other only by the type of pigments used (Table 1). The pigmented coatings (O1, O2, AL2, S1) were characterized by significantly lower colour changes than transparent ones.

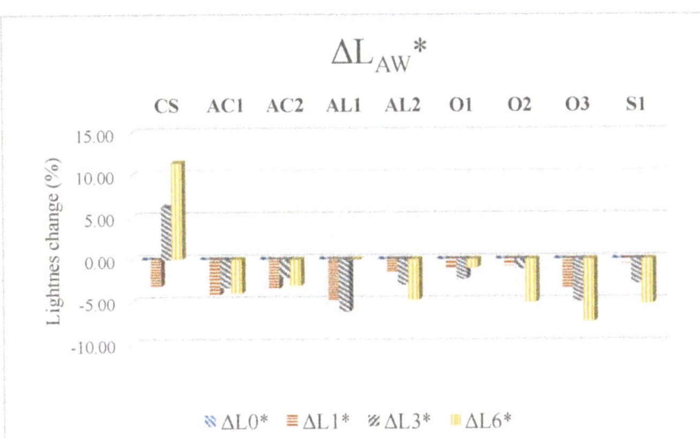

Figure 1. *Cont.*

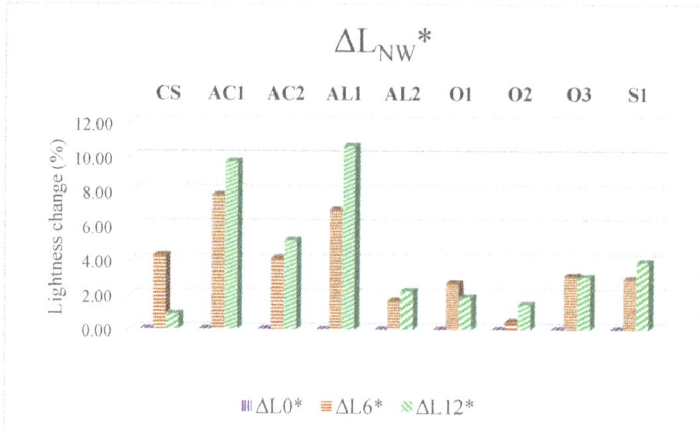

Figure 1. Lightness difference (ΔL^*) of tested coating systems during artificial and natural weathering (CS means control sample).

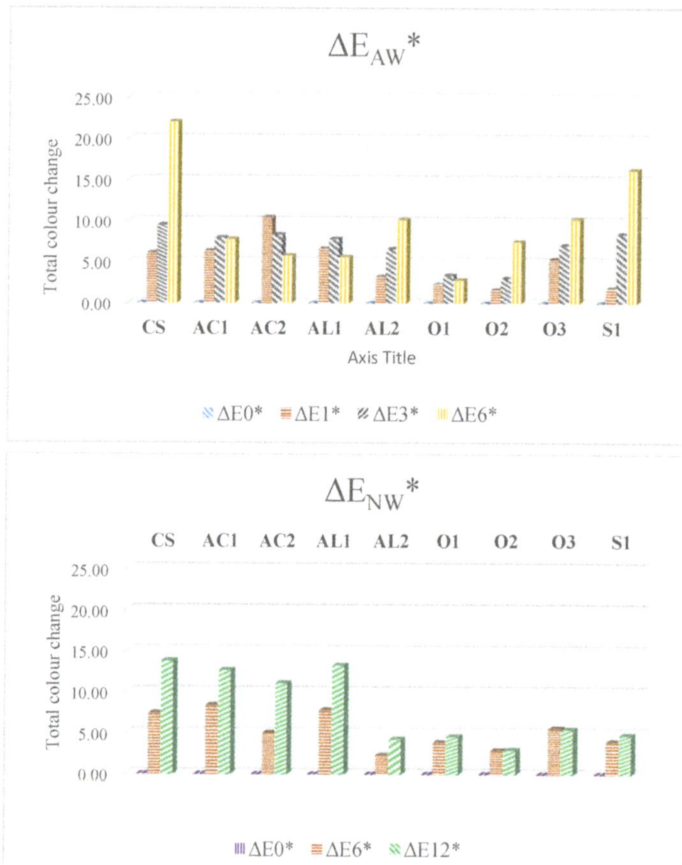

Figure 2. Total colour difference (ΔE^*) of tested coating systems during artificial and natural weathering (CS means control sample).

3.2. Gloss Change of the Samples

Except for small fluctuations, all coatings were characterized by decreased gloss values during both weathering methods (Figure 3). The reference samples, on the other hand, manifested increases in this property (with the decrease in the final phase of NW). The same findings regarding protected and unprotected weathered wood were found by Ghosh et al. [46]. The best results were noted for acrylate and AL1 coatings, while the highest changes were recorded for oil coatings.

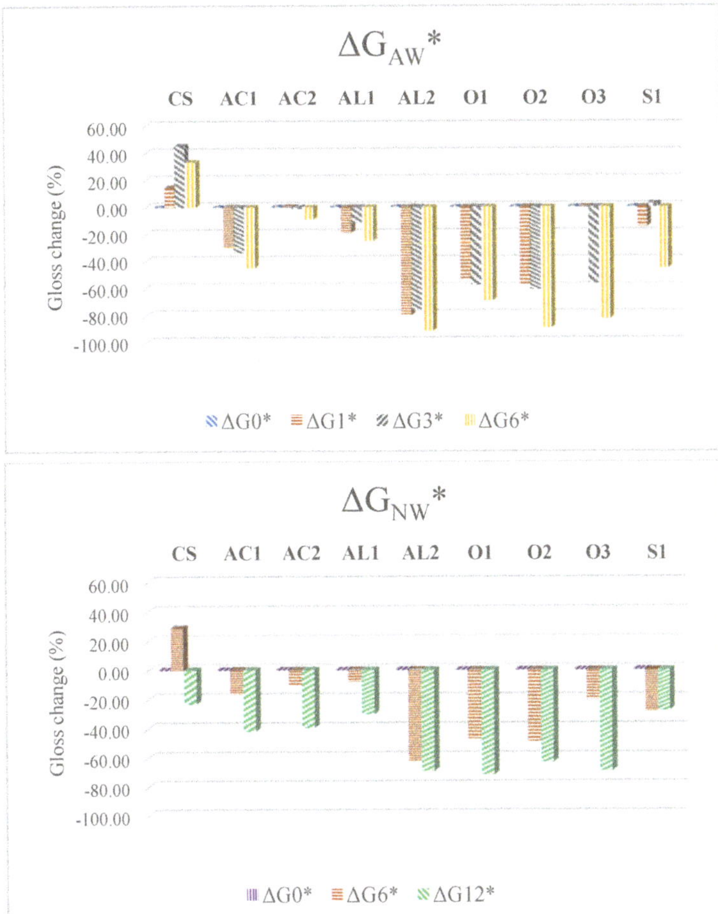

Figure 3. Gloss difference (ΔG*) of tested coating systems during artificial and natural weathering (CS means control sample).

3.3. Surface Wettability of the Samples

The contact angle, which indicates the wettability by water on the exposed surfaces of coated wood, is an important indicator of the rate of weathering [22,24]. Surface wetting changes (Figure 4) indicated the overall impairment of the protective function of the coating systems against water [24,39]. During AW, the most stable values were noted for acrylate coating systems and O3. The rest of the samples were characterized by decreased surface wettability (the most in the case of reference samples and AL2). During NW, the wettability decreased for all samples, but the smallest changes were also recorded for acrylate and O3 coating.

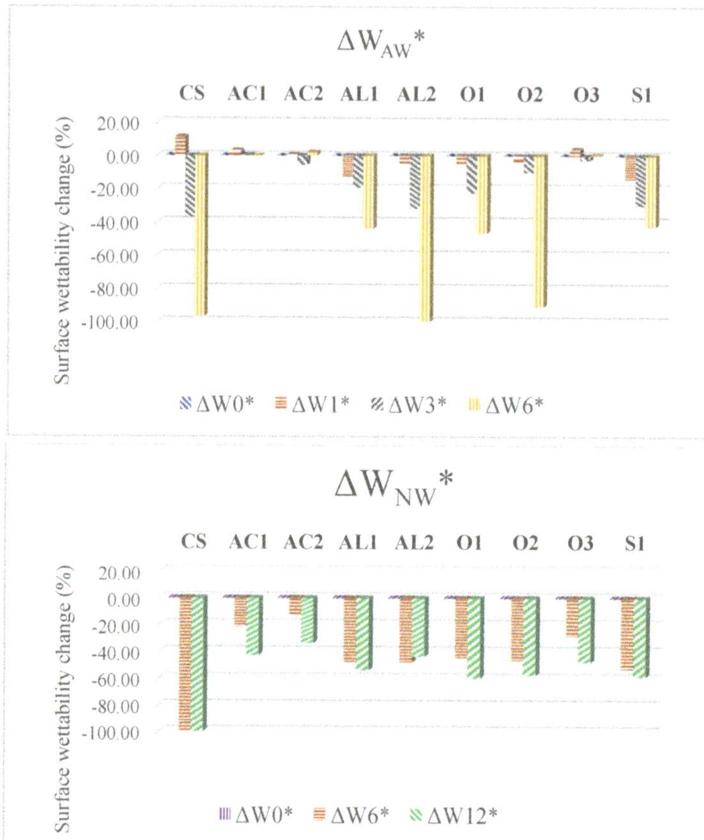

Figure 4. Surface wettability difference (ΔW*) of the tested coating systems during artificial and natural weathering (CS means control sample).

The Spearman rank correlation of properties after 12 months of NW in comparison with 6 weeks of AW measured both the strength and direction of the relationship between the ranks of data (Table 4).

Table 4. Spearman rank correlation between AW and NW.

	Type of Coating	Number of Valid Tests	Spearman R	p-Value
ΔE_{6AW} * × ΔE_{12NW} *	Transparent	5	0.10	0.87
	Pigmented	4	0.40	0.60
	All coatings	9	0.18	0.64
ΔL_{6AW} * × ΔL_{12NW} *	Transparent	5	−0.10	0.87
	Pigmented	4	−0.40	0.60
	All coatings	9	−0.03	0.93
ΔG_{6AW} * × ΔG_{12NW} *	Transparent	5	0.90	0.04 *
	Pigmented	4	0.40	0.60
	All coatings	9	0.77	0.02 *
ΔW_{6AW} * × ΔW_{12NW} *	Transparent	5	0.90	0.04 *
	Pigmented	4	−0.80	0.20
	All coatings	9	0.53	0.14

Note: * means statically significant at 95% level ($p < 0.05$); R = 1 is a perfect positive correlation; R = −1 is a perfect negative correlation; R = 0 is no correlation.

Based on the results of Spearman rank correlation, strong statistically significant relationships between results from AW and NW were only found with gloss changes of all coatings and surface wettability changes and gloss changes of transparent coatings ($p < 0.05$). Further evaluation revealed the remaining results from AW and NW, including total colour difference were statistically insignificant ($p > 0.05$) and very poorly correlated with each other.

3.4. Macroscopic and Microscopic Evaluation of the Samples

The coating performance of the samples during NW was also evaluated visually in accordance with other studies [18,20]. The visual evaluation confirmed that weathering causes colour changes and surface degradation both in natural and laboratory conditions [23,24]. Visual inspection (Figure 5 for AW and Figure 6 for NW) confirmed the exact previously measured values–darkening of coated samples during AW and lightening of samples during NW (Figure 1) and associated total colour (ΔE^*) and gloss changes (Figures 2 and 3).

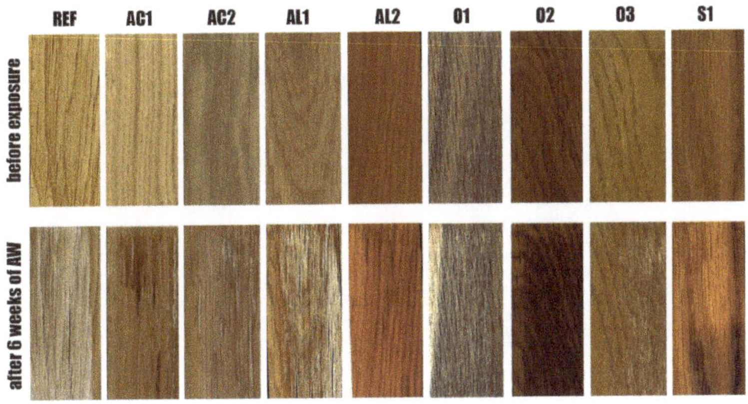

Figure 5. Visual evaluation of tested coatings before and after 6 weeks of AW exposure.

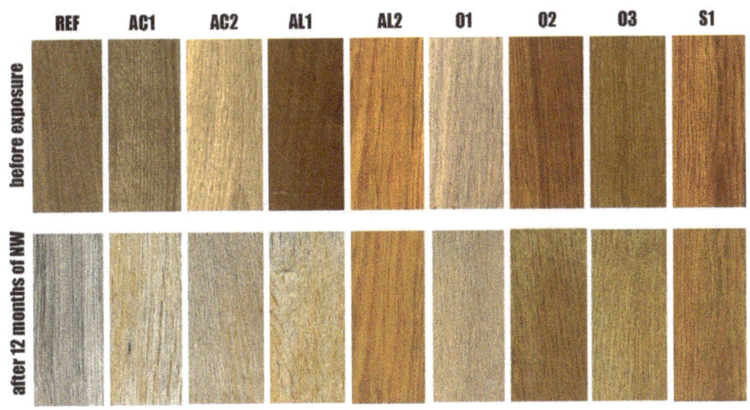

Figure 6. Visual evaluation of tested coatings before and after 12 months of NW exposure.

Confocal laser scanning microscopy was employed to assess degradation of selected coatings. Figure 7 illustrates the degradation of oil coating (O1) and acrylate coating (AC1) after 6 weeks of AW or 12 months of NW. Lower colour changes were noted for oil coatings (Figure 2), but also a more obvious disruption and degradation of the coating surface (Figure 7) which are more connected with the higher gloss and surface wettability changes. In the line with this, the acrylate coatings were characterized by lower gloss and surface wettability changes associated with the lower degree of coating degradation.

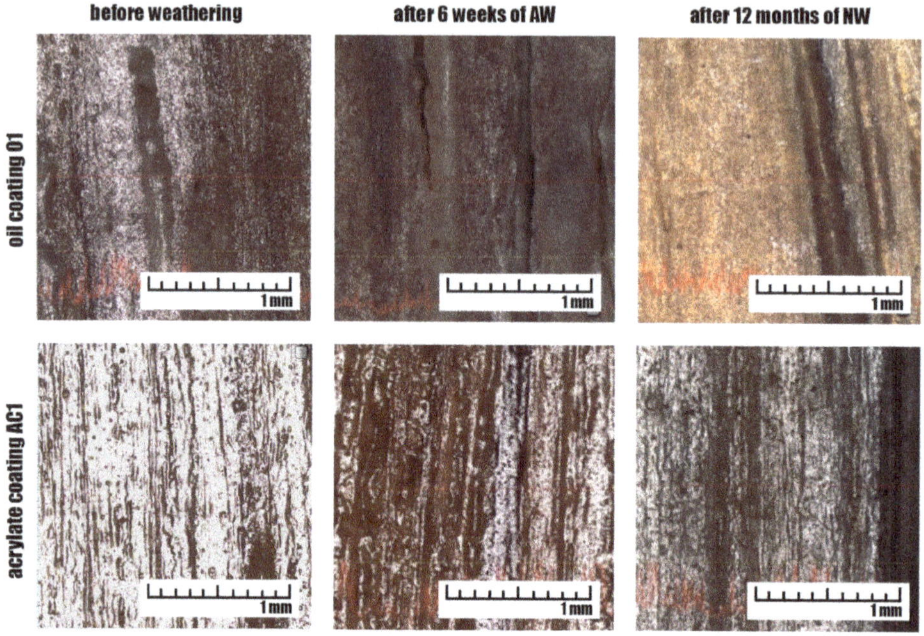

Figure 7. Confocal microscopy of oil (O1) and acrylate (AC1) coating after 6 weeks of AW and 12 months of NW.

3D images of samples roughness profile for selected tested coatings with higher gloss change (AL2 and O3) are shown in Figure 8. It is possible to see increasing of roughness of the surfaces after 6 weeks of accelerated weathering. These images (Figure 8) also confirm that decreasing of gloss is influenced mainly by decomposition of coating film (O3) or its top layer (AL2) (see also Figure 5).

Native oak wood has higher natural durability and lower colour changes in comparison with other hardwood species during weathering [47], but to find durable coatings systems suitable for its specific chemical and morphological structure is desirable. The findings of this study confirmed the effect of polymer base on the overall performance of coatings [12,48]. It is clear that effect of surface protection against weathering was demonstrated by the difference between uncoated reference and coated samples (Figures 1–6). Generally, the oil coatings (O1–O3) performed well in the colour analyses, the acrylate coatings (AC1–AC2) reached the best results in the gloss and wettability evaluation. These properties are more likely connected with coating degradation and disruption than with chemical changes in wood [34,39]. Acrylate and oil coatings reached the best performance on larch wood also in study of Šimůnková et al. [15]. In opposite, in the study of Sivrikaya et al. [48], the better performance against atmospheric conditions on oak wood was recorded for alkyd coatings compared with other tested coatings.

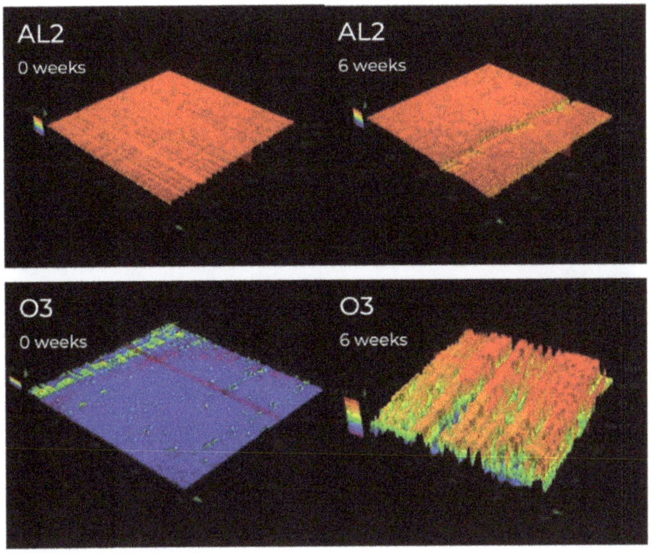

Figure 8. 3D images of surfaces before (left) and after 6 weeks of artificial accelerated weathering (right). Figures were created using Confocal laser scanning microscope with 108-fold magnification. Size of analysed area is 2560 μm × 2560 μm.

The total colour changes (ΔE*) were the most connected with the change of lightness parameter ΔL* (Figures 1 and 2) as in the other studies [43,49]. The lowest colour changes were observed for oil coatings O1 and O2 (thin-layer oil-based with micronized pigments TiO_2 and Fe_2O_3). The positive effect of white TiO_2 pigments on photodegradation was already discussed in the work of Moya et al. [30]. The thickness of the coating system is a criterion affecting its service life [23] but for tested coatings in this work, the pigment content was the more important factor (Table 1, Figures 5 and 6). The pigmented coatings generally provided more effective protection and reached the lower colour changes during NW than transparent ones as in the study of Sivrikaya et al. [48].

In the gloss and surface wettability evaluation of the samples, the performance of coating differed in comparison with colour analysis. All the tested samples except control samples were noted for loss of gloss during NW and AW. The best results were observed for acrylate AC1-AC2 and AL1 coatings. Although the role of gloss change is still discussed - according to Pánek et al. [39], the gloss change is more sensitive to the coating degradation than to total colour difference (Figure 8). Merlatti et al. [34] stated that loss of gloss should not be systematically correlated to the advance in chemical degradation during weathering. The lowest change of surface wettability was recorded for acrylate coating systems AC1-AC2 and oil coating O3 both after AW and NW. The rest of the samples were characterized by decrease of surface wettability, this was the most significant in the case of control and AL2 samples.

The comparison of both weathering methods only by evaluating colour difference would be insufficient, as stated in other previously done studies [23,28,30]. The combination of different testing parameters of coating systems and visual evaluation gives a better idea about the durability of coatings [49,50], despite that the total colour difference still remains the main indicator of coating degradation. Coatings defoliated after AW were in most cases highly degraded after 12 months of NW (Figure 5 versus Figure 6). Based on this, AW can be recommended as the first step for selection of nondurable coatings mainly on woods with specific chemical or morphological structure. The non-linear correlations were performed to compare the strength of the relationship between the total colour differences, gloss and surface after NW and AW of transparent and pigmented coatings as in the study of Pánek and Reinprecht [51]. The results were highly varied, and, in the most cases, without any statistical significance. Comparisons of colour changes mainly showed weak correlation for tested oak, as for black locust and spruce wood in the work Pánek and Reinprecht [52]. The strong correlation was found only for the gloss changes of all coatings, in agreement with the work Q-Lab [53], and surface wettability and gloss change of transparent coatings separately. Both methods revealed the certain durability among the tested coating systems and come to the greater agreement than in the case of unprotected wood weathering. These inconsistent results confirm the previously stated difficulty to mathematically correlate data from outdoor and laboratory conditions [22,30]. The significant impact of climatic and local environmental conditions at the testing site is still one of the dominant factors preventing the accurate prediction of real weathering in the exterior via artificial accelerated weathering [36,54]. Even there is an effort to simulate outdoor conditions in UV chamber by setting the parameters of weathering, the more accurate correlation for prediction changes of coated wood via artificial weathering in laboratory has proven to be difficult. Despite the obvious advantages of artificial weathering, the results provided by this method still lack reliability of natural weathering and they should always be carefully interpreted and in the best scenario accompanied by natural weathering tests to verify the performance of coatings in an end-use environment [11].

4. Conclusions

Eight different transparent and pigmented coating systems were applied on oak samples and tested using artificial and natural weathering. Total colour difference was mostly related to the lightness parameter change. Evaluation of discolouration with other criteria such as gloss, surface wettability and visual and microscopic evaluation more accurate predicted service life of coatings. Pigmented coatings had significantly lower colour changes than transparent ones, for both artificial and natural weathering. Oil coatings were more colour stable, acrylate coatings achieved the best results of gloss and wettability changes. The gloss and surface wettability changes better copy degradation and disruption of coated wood in comparison with total colour changes. Even there were some visually observed similarities in the test results of AW and NW exposed samples, this was not confirmed statistically. The Spearman rank correlation showed strong statistically significant relationship between results after artificial and natural weathering only for the gloss changes of all coatings and surface wettability and gloss changes of transparent coatings separately.

Author Contributions: Conceptualization and Methodology M.P.; Formal Analysis, E.O.; Investigation, K.Š., O.D., I.Š.; Writing—Original Draft Preparation, E.O.; Writing—Review & Editing, E.O., M.P., S.H.; Supervision, M.P.; Project Administration, P.Š.

Funding: This work was supported by the grant "Improving of coatings durability on selected kinds of wood in the exterior applications," No. TH02020873, financed by TA ČR.

Conflicts of Interest: The authors declare no conflict of interest.

References

1. EN 350. *Durability of Wood and Wood-Based Products—Testing and Classification of the Durability to Biological Agents of Wood and Wood-Based Materials*; European Committee for Standardization: Brussels, Belgium, 2016.
2. Zahri, S.; Belloncle, C.; Charrier, F.; Pardon, P.; Quideau, S.; Charrier, B. UV light impact on ellagitannins and wood surface colour of European oak (*Quercus petraea* and *Quercus robur*). *Appl. Surf. Sci.* **2007**, *253*, 4985–4989. [CrossRef]
3. Sivrikaya, H. Impregnability and Durability Characteristics of Sapwood and Heartwood. Ph.D. Thesis, Zonguldak Karaelmas University, Graduate School of Natural and Applied Sciences, Zonguldak, Turkey, 2003.
4. Ayadi, N. Vieillissement Climatique d'un Systéme Bois-Vernis-Absorbeur UV Inorganique. Ph.D. Thesis, Université de Nantes, Nantes, France, 2004.
5. Browne, F.L. *Wood Properties that Affect Paint Performance*; U.S. Forest Service, Forest Products Laboratory: Madison, WI, USA, 1958.
6. Oberhofnerová, E.; Pánek, M.; García-Cimarras, A. The effect of natural weathering on untreated wood surface. *Maderas Ciencia y Tecnología* **2017**, *19*, 173–184. [CrossRef]

7. George, B.; Suttie, E.; Merlin, A.; Deglise, X. Photodegradation and photostabilisation of wood—The state of the art. *Polym. Degrad. Stab.* **2005**, *88*, 268–274. [CrossRef]
8. Auclair, N.; Riedl, B.; Blanchard, V.; Blanchet, P. Improvement of photoprotection of wood coatings by using inorganic nanoparticles as ultraviolet absorbers. *For. Prod. J.* **2011**, *61*, 20–27. [CrossRef]
9. Feist, W.C.; Hon, D.N.S. Chemistry of weathering and protection. In *The Chemistry of Solid Wood*; ACS: Washington, DC, USA, 1984; pp. 401–451.
10. Evans, P.D. Weathering and photoprotection of wood. In *Development of Commercial Wood Preservatives*; ACS: Washington, DC, USA, 2008; Volume 982, pp. 69–117. [CrossRef]
11. Cogulet, A.; Blanchet, P.; Landry, V. Evaluation of the impacts of four weathering methods on two acrylic paints: Showcasing distinctions and particularities. *Coatings* **2019**, *9*, 121. [CrossRef]
12. De Meijer, M. Review on the durability of exterior wood coatings with reduced VOC-content. *Prog. Org. Coat.* **2001**, *43*, 217–225. [CrossRef]
13. Evans, P.; Haase, J.; Seman, A.S.B.M.; Kiguchi, M. The search for durable exterior clear coatings for wood. *Coatings* **2015**, *5*, 830–864. [CrossRef]
14. Miniutti, V.P. Microscale changes in cell structure at softwood surfaces during weathering. *For. Prod. J.* **1964**, *14*, 571–576.
15. Šimůnková, K.; Oberhofnerová, E.; Reinprecht, L.; Pánek, M.; Podlena, M.; Štěrbová, I. Durability of selected transparent and semi-transparent coatings on Siberian and European larch during artificial weathering. *Coatings* **2019**, *9*, 39. [CrossRef]
16. Forsthuber, B.; Ecker, M.; Truskaller, M.; Grüll, G. Rapid prediction of surface characteristics of European and Siberian larch wood by FT-NIRS. *Eur. J. Wood Wood Prod.* **2017**, *75*, 569–580. [CrossRef]
17. Gobakken, L.R.; Lebow, P.K. Modelling mould growth on coated modified and unmodified wood substrates exposed outdoors. *Wood Sci. Technol.* **2010**, *44*, 315–333. [CrossRef]
18. De Windt, I.; Van den Bulcke, J.; Wuijtens, I.; Coppens, H.; Van Acker, J. Outdoor weathering performance parameters of exterior wood coating systems on tropical hardwood substrates. *Eur. J. Wood Wood Prod.* **2014**, *72*, 261–272. [CrossRef]
19. Gaylarde, C.C.; Morton, L.H.G.; Loh, K.; Shirakawa, M.A. Biodeterioration of external architectural paint films—A review. *Int. Biodeterior. Biodegrad.* **2011**, *65*, 1189–1198. [CrossRef]
20. Grüll, G.; Truskaller, M.; Podgorski, L.; Bollmus, S.; Tscherne, F. Maintenance procedures and definition of limit states for exterior wood coatings. *Eur. J. Wood Wood Prod.* **2011**, *69*, 443–450. [CrossRef]
21. Dawson, B.S.; Singh, A.P.; Kroese, H.W.; Schwitzer, M.A.; Gallagher, S.; Riddiough, S.J.; Wu, S. Enhancing exterior performance of clear coatings through photostabilization of wood. Part 2: Coating and weathering performance. *J. Coat. Technol. Res.* **2008**, *5*, 207–219. [CrossRef]
22. Van den Bulcke, J.; De Windt, I.; Defoirdt, N.; De Smet, J.; Van Acker, J. Moisture dynamics and fungal susceptibility of plywood. *Int. Biodeterior. Biodegrad.* **2011**, *65*, 708–716. [CrossRef]
23. Grüll, G.; Tscherne, F.; Spitaler, I.; Forsthuber, B. Comparison of wood coating durability in natural weathering and artificial weathering using fluorescent UV-lamps and water. *Eur. J. Wood Wood Prod.* **2014**, *72*, 367–376. [CrossRef]
24. Žlahtič, M.; Humar, M. Influence of artificial and natural weathering on the hydrophobicity and surface properties of wood. *BioResources* **2016**, *11*, 4964–4989. [CrossRef]
25. Miklečić, J.; Turkulin, H.; Jirouš-Rajković, V. Weathering performance of surface of thermally modified wood finished with nanoparticles-modified waterborne polyacrylate coatings. *Appl. Surf. Sci.* **2017**, *408*, 103–109. [CrossRef]
26. EN 927-3. *Paints and Varnishes. Coating Materials and Coating System for Exterior Wood, Part 3: Natural Weathering Test*; European Committee for Standardization: Brussels, Belgium, 2006.
27. EN 927-6. *Paints and Varnishes. Coating Materials and Coating Systems for Exterior Wood, Part 6: Exposure of Wood Coatings to Artificial Weathering Using Fluorescent UV Lamps and Water*; European Committee for Standardization: Brussels, Belgium, 2006.
28. Reinprecht, L.; Pánek, M. Effects of wood roughness, light pigments, and water repellent on the color stability of painted spruce subjected to natural and accelerated weathering. *BioResources* **2015**, *10*, 7203–7219. [CrossRef]
29. Deflorian, F.; Rossi, S.; Fedrizzi, L.; Zanella, C. Comparison of organic coating accelerated tests and natural weathering considering meteorological data. *Prog. Org. Coat.* **2007**, *59*, 244–250. [CrossRef]

30. Moya, R.; Rodríguez-Zúñiga, A.; Vega-Baudrit, J.; Puente-Urbina, A. Effects of adding TiO_2 nanoparticles to a water-based varnish for wood applied to nine tropical woods of Costa Rica exposed to natural and accelerated weathering. *J. Coat. Technol. Res.* **2017**, *14*, 141–152. [CrossRef]
31. Oberhofnerová, E.; Pánek, M.; Böhm, M. Effect of surface pretreatment with natural essential oils on the weathering performance of spruce wood. *BioResources* **2018**, *13*, 7053–7070. [CrossRef]
32. Valverde, J.C.; Moya, R. Correlation and modeling between color variation and quality of the surface between accelerated and natural tropical weathering in Acacia mangium, Cedrela odorata and Tectona grandis wood with two coating. *Color Res. Appl.* **2014**, *39*, 519–529. [CrossRef]
33. Jirous-Rajkovic, V.; Bogner, A.; Radovan, D. The efficiency of various treatments in protecting wood surfaces against weathering. *Surf. Coat. Int. B Coat. Trans.* **2004**, *87*, 15–19. [CrossRef]
34. Merlatti, C.; Perrin, F.X.; Aragon, E.; Margaillan, A. Natural and artificial weathering characteristics of stabilized acrylic–urethane paints. *Polym. Degrad. Stab.* **2008**, *93*, 896–903. [CrossRef]
35. Oltean, L.; Teischinger, A.; Hansmann, C. Wood surface discolouration due to simulated indoor sunlight exposure. *Holz Roh-und Werkst.* **2008**, *66*, 51. [CrossRef]
36. Creemers, J.; De Meijer, M.; Zimmermann, T.; Sell, J. Influence of climatic factors on the weathering of coated wood. *Eur. J. Wood Wood Prod.* **2002**, *60*, 411–420. [CrossRef]
37. ČSN 49 0108. *Drevo. Zist'ovanie Hustoty [Wood. Determination of the Density]*; Český Normalizační Institut: Prague, Czech Republic, 1993.
38. Meteostation of the Faculty of Agronomy—Department of Agroecology and Biometeorology, Prague, Czech Republic. 2019. Available online: http://meteostanice.agrobiologie.cz (accessed on 27 November 2019).
39. Pánek, M.; Oberhofnerová, E.; Zeidler, A.; Šedivka, P. Efficacy of hydrophobic coatings in protecting oak wood surfaces during accelerated weathering. *Coatings* **2017**, *7*, 172. [CrossRef]
40. Commission Internationale de l'Eclairage (CIE). *Colorimetry*, 2nd ed.; (CIE Pub. No. 15.2); Commission Internationale de l'Eclairage: Vienna, Austria, 1986.
41. EN ISO 2813. *Paints and Varnishes—Determination of Gloss Value at 20 Degrees, 60 Degrees and 85 Degrees*; European Committee for Standardization: Brussels, Belgium, 2014.
42. Bastani, A.; Adamopoulos, S.; Militz, H. Water uptake and wetting behaviour of furfurylated, N-methylol melamine modified and heat-treated wood. *Eur. J. Wood Wood Prod.* **2015**, *73*, 627–634. [CrossRef]
43. Ozgenc, O.; Hiziroglu, S.; Yildiz, U.C. Weathering properties of wood species treated with different coating applications. *BioResources* **2012**, *7*, 4875–4888. [CrossRef]
44. Turkoglu, T.; Baysal, E.; Toker, H. The effects of natural weathering on color stability of impregnated and varnished wood materials. *Adv. Mater. Sci. Eng.* **2015**. [CrossRef]
45. Sehlstedt-Persson, M. Color responses to heat-treatment of extractives and sap from pine and spruce. In Proceedings of the 8th IUFRO International Wood Drying Conference: Improvement and Innovation in Wood Drying: A Major Issue for a Renewable Material, Brasov, Romania, 24–29 August 2003; Faculty of Wood Industry, Transilvania University of Brasov: Brasov, Romania, 2003; pp. 459–464.
46. Ghosh, M.; Gupta, S.; Kumar, V.S. Studies on the loss of gloss of shellac and polyurethane finishes exposed to UV. *Maderas Ciencia y Tecnología* **2015**, *17*, 39–44. [CrossRef]
47. Kubovský, I.; Oberhofnerová, E.; Kačík, F.; Pánek, M. Surface changes of selected hardwoods due to weather conditions. *Forests* **2018**, *9*, 557. [CrossRef]
48. Sivrikaya, H.; Hafizoglu, H.; Yasav, A.; Aydemir, D. Natural weathering of oak (*Quercus petrae*) and chestnut (*Castanea sativa*) coated with various finishes. *Color Res. Appl.* **2011**, *36*, 72–78. [CrossRef]
49. Temiz, A.; Terziev, N.; Eikenes, M.; Hafren, J. Effect of accelerated weathering on surface chemistry of modified wood. *Appl. Surf. Sci.* **2007**, *253*, 5355–5362. [CrossRef]
50. Singh, T.; Singh, A.P. A review on natural products as wood protectant. *Wood Sci. Technol.* **2012**, *46*, 851–870. [CrossRef]
51. Pánek, M.; Reinprecht, L. Colour stability and surface defects of naturally aged wood treated with transparent paints for exterior constructions. *Wood Res.* **2014**, *59*, 421–430.
52. Pánek, M.; Reinprecht, L. Critical view on the possibility of color changes prediction in the surfaces of painted wood exposed outdoors using accelerated weathering in Xenotest. *J. Coat. Technol. Res.* **2019**, *16*, 339–352. [CrossRef]

53. Q-LAB. Correlation of Laboratory to Natural Weathering. Technical Bulletin LU-0824. 1977. Available online: www.q-lab.com/resources/technical-articles.aspx (accessed on 16 December 2019).
54. Mattos, B.D.; De Cademartori, P.H.G.; Lourençon, T.V.; Gatto, D.A. Colour changes of Brazilian eucalypts wood by natural weathering. *Int. Wood Prod. J.* **2014**, *5*, 33–38. [CrossRef]

coatings

MDPI

Article

Wettability of Wood Surface Layer Examined From Chemical Change Perspective

Eva Annamaria Papp [1], Csilla Csiha [1], Adam Nandor Makk [2], Tamas Hofmann [2] and Levente Csoka [3,*]

[1] Institute of Wood Based Products and Technologies, University of Sopron, Bajcsy-Zs. E. 4, 9400 Sopron, Hungary; papp.eva.annamaria@uni-sopron.hu (E.A.P.); csiha.csilla@uni-sopron.hu (C.C.)

[2] Institute of Chemistry, University of Sopron, Bajcsy-Zs. E. 4, 9400 Sopron, Hungary; makk.adam.nandor@uni-sopron.hu (A.N.M.); hofmann.tamas@uni-sopron.hu (T.H.)

[3] Institute of Cellulose and Paper Technology, celltech-paper Ltd., 9400 Sopron, Hungary

* Correspondence: csoka.levente@celltech-paper.hu

Received: 17 February 2020; Accepted: 8 March 2020; Published: 10 March 2020

Abstract: The effect of artificial ageing on spruce (*Picea abies*), beech (*Fagus sylvatica* L.), birch (*Betula pendula*), and sessile oak (*Quercus petraea*) wood surfaces were investigated using qualitative (total phenolic and total soluble carbohydrate content) chemical examination methods. During ageing (∑240h), the influence of surface chemistry modifications was monitored by contact angle measurements of polar, dispersive (distilled water), and dispersive (diiodomethane) liquids. The results clearly show the relation between the ratio of main chemical components of the wood surface layer and surface wettability during artificial radiation. The identified surface chemistry modifications cause more significant change in the contact angle of polar and dispersive liquid, relative to the change of dispersive liquid contact angle. Chemical changes of the wood surface layer are due to the degradation of the main wood components (cellulose, hemicelluloses, and lignin) which can be properly monitored by total phenolic (TPC) and total soluble carbohydrate content (TSCC) measurements.

Keywords: wettability; phenol; carbohydrates; beech; birch; spruce; sessile oak

1. Introduction

Regarding the antioxidant capacity of wood extracts, including phenolic compounds, numerous studies focus on the total phenolic content measurements in food research or medical biology [1]. Only a few researches deal with the investigation of extractive content of wood, while most researches report on the mechanical properties of wood. The main goal of this research was to detect the relation between changes of chemical components and the apparent contact angle of wood surface during long-term artificial radiation. To find these relations, total carbohydrate and total phenolic compound measurements were performed in parallel with contact angle measurements to identify decomposition products of the wood surface thin layer and to monitor the behavior of contact angle of the wood surface.

There are considerable differences between wood species regarding the ratio of lignin, as their structural material. While the lignin content of hardwoods is 18%–25%, for softwoods, it is between 25% and 35% generally [2]. The considerable diversity in quality and quantity of wood components is reflected in the amount of extractives. There can be significant differences in extractive content of samples coming from different species [3]. Although extractives are present only in minor quantity relative to the quantity of main chemical wood components, their influence is intense on the chemical nature of different wood species. Extractives affect the color and odor of wood, while even the efficiency of such different processing technologies as pulping and drying have and impact on the durability, adhesion, and hygroscopic behavior of wood, as noted by Umezawa [4]. Previous researches described the relation between wood mechanical properties and phenolic compounds. Aloui et al. [5] demonstrated the effect of phenolic compounds on the durability of oak wood and defined that the higher the proportions of phenol, the higher the resistance of wood is. According to Trapp et al. [6], in terms of wood surface, the intensification of hydrophobic character can be primarily related to the increase of phenolic compounds quantity. The high degree of wood wettability can be attributed to the various hydrophilic components (e.g., hemicelluloses) of the wood surface [7]. Many studies suggest that adhesion is highly affected by extractives of wood and the wood surface layer [7–11]. High concentration extract material on wood surfaces could become a physical obstacle, making difficult the chemical bonding between, e.g., glue and wood. This type of extractive layer can decrease the surface energy and extent of wettability, and can degrade the penetration of glue into the wood material [12]. According to certain studies, there isn't a clear correlation between adhesion and the quantity of extractives [13], however Nguyen and Johns [14] stated that the percentage of wood extractive content is in inverse proportion to wettability. During contact angle measurements, wood species having the higher extractive and lignin content were found to have the higher contact angle values [15–17]. The different parts of wood due to their chemical and structural complexity can be wetted in different manner and with different success, typically decreasing with the increase of the extract materials [18]. The quantitative difference of extractives is able to cause even 40% variance with regard to wettability [19].

The wettability of wood depends on several factors due to its complexity, such as species, storage conditions (water, sunlight, biotic and abiotic factors), drying process, and cutting direction, as stated by Nguyen and Johns [20] and Kalnins et al. [21]. Wettability is also affected by the chemical composition of the surface, as stated by Kishino and Nakano [22], and the density of wood, as noticed by Amorim et al. [23].

One of the simplest methods to determine the surface energy of a solid surface is to measure the contact angle of different liquid drops on the wood surface [24]. Theoretically, the contact angle instantaneously characterizes the feature of the solid-liquid system and is measurable in a given condition [25]. Assuming that instantaneous contact angle changes are consequence of hydrodynamic processes, it could be stated hypothetically that its most characteristic value can be measured in that moment, when the shape of the liquid droplet is formed on the surface (or at a very close time) [26]. Behavior of a liquid drop relaxing on a solid surface are influenced by the changes of the wood surface [27]. The reason for this is that the contact angle is a phenomenological parameter which is not influenced just by surface energy, but by surface roughness, surface heterogeneity, moisture content, and a lot of other factors of the wood material [28].

The photodegradation of wood makes it more complicated to find the influencing factors of wettability. Those research results are relevant for us which deal with the chemical changes of wood surface caused by artificial ageing, and with the influence of radiation-caused degradation and its effect on contact angle.

From the adhesion point of view, one of the most critical factors is the time elapsed since machining. In the case of wood, the most optimal surfaces for gluing and surface treatments are freshly prepared surfaces [9,29]. Accordingly, the wettability of different wood species decreases in parallel with the age of machined surfaces because of the chemical transformation of extractives on wood surfaces [30]. Wood surfaces show significant changes in their surface free energy even after days of conditioning, as reported in different studies [31,32]. Lignin is easily oxidisable by photodegradation and its structural changes can be detected right when it comes in contact with the air, or during long-term storage [33]. Surface changes of wood due to natural aging can be imitated using artificial ageing apparatus under laboratory conditions. Both processes can be blamed for decreasing the surface energy parameter, caused by the migration of hydrophobic extractives onto the surface of wood [34].

It is noted that lignin together with extractives is able to increase the hydrophobic character of wood surface. Unlike the cellulose and hemicellulose, lignin has relatively hydrophobic character [35]. Lignin is able to play a role in the increase of hydrophobicity, being active in moisture transport in wood [36,37]. Williams et al. [38] stated that light irradiation causes rapid colour changes followed by processes having a strong impact on surface wettability in the case of some wood species. Similarly, to other natural polymers lignin and polyphenols, as wood components, absorb UV radiation. Due to UV absorption, photolytic, photo-oxidative, and thermo-oxidative reactions start to develop in wood material and those reactions are responsible for the degradation caused by sunlight [39]. Radicals formed due to the photodegradation of lignin protect lignin and also the complex wood system from further photodegradation effects [40]. Besides lignin, the most intensive visible changes due to photodegradation are caused by extractives. The reason of this is that extractives have strong light absorption attribute [41,42] and in this manner protect the main wood components [43]. During artificial or natural ageing hydrophobic extractives migrate to the wood surface and cause decrease of the wettability [34]. The dissolution of phenolic compounds is result of surface structure also, which is typical when the wood surface comes into contact with liquids [44–47]. In case of some wood species due to artificial ageing wettability increases. The phenomenon of ablation caused by UV radiation is the reason for the mitigation of materials able to decrease wettability of wood surface [48]. Due to long-term UV radiation of the cell wall, researchers [49] detected significant quantity of water-soluble decomposition products, which can alter the quality of wettability.

Penetration depth of UV radiation and visible light are different, and the photodegradation caused by them is of different measure also. Based on Williams [50] and Pandey [41] the usual penetration depth of UV radiation is ~75 μm. These values depend on the density of wood and ratio of earlywood and latewood. Hon and Ifju [51] stated that the maximum UV penetration depth in wood is not more than 80 μm. According to the research of Németh and Faix [52], the upper 75-μm thin layer of wood should be monitored when examining photodegraded wood surfaces.

The present investigations are directed towards the development of a novel method for the evaluation of wood surface chemical compound changes under artificial ageing in the case of different wood species.

2. Materials and Methods

During the investigation of artificially aged wood surfaces, contact angle measurements (using distilled water and diiodomethane), moisture content (MC) examinations, total phenolic and total soluble carbohydrate content measurements were performed on wood samples of beech (*Fagus sylvatica* L.), birch (*Betula pendula*), sessile oak (*Quercus petraea*), and spruce (*Picea abies*), as depicted in Figure 1.

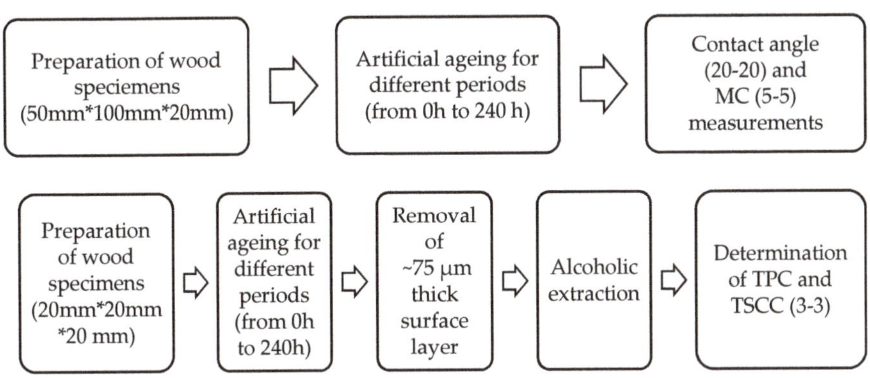

Figure 1. Flow diagram of experimental procedures.

2.1. *Preparation of Wood Specimens*

Before ageing, all wood surfaces were sanded with sandpaper of grit size 150. From the different wood species similar samples were prepared for different measurements, including the specimens for contact angle and moisture content examinations from different boards with dimensions

$50 \times 100 \times 20$ mm^3 and specimens for total phenol and total carbohydrate content measurements with dimensions $20 \times 20 \times 20$ mm^3.

2.2. Technological Background of Artificial Ageing

The artificial ageing of different wood species was performed with Original Hanau Suntest (HERAEUS, Hanau, Germany) equipped with xenon bulb and built-in UV mirror, started within 1 h after machining. The different measurements were performed in the following ageing periods: 0 h (control), 1 h, 3 h, 5 h, 8 h, 10 h, 15 h, 20 h, 30 h, 60 h, 96 h, 132 h, 174 h, and 240 h, respectively. During artificial ageing, the temperature of wood surfaces was monitored by using Maxwell MX 25-903 type (GLOBIZ, Ashford, Kent, UK) digital thermometer. Temperature monitoring was responsible for the chamber temperature do not exceed the 55 ± 5 °C value which was defined as maximum reachable ageing temperature. Based on practical experience of artificial ageing the temperature over 60 °C can induce major surface chemical changes.

For moisture content measurements HM8-WS5 Merlin Moisture Meter (MERLIN Technology GmbH, Tumeltsham, Austria) was used, with ~5 mm scanning depth, and ~20 cm^2 maximum measuring area to perform 5 moisture content examinations on each wood surface after ending different ageing periods.

2.3. Contact Angle Measurements of Artificially Aged Wood Surfaces

Dynamic contact angle measurements were performed on each wood surface by using PG-X goniometer (FIBRO SYSTEMS AG, Hagersten, Sweden): 20 by using distilled water and 20 by using diiodomethane (SIGMA-ALDRICH, St. Louis, MO, USA). Measurements were performed in case of all ageing periods, 1 sec after the drop release and formation of the liquid droplet on the surface, as previously agreed [26].

2.4. Alcoholic Extraction and Determination of Total Phenolic and Total Soluble Carbohydrate Contents

To measure the total phenolic and total soluble carbohydrate contents, a ~75-µm thick layer was collected by alcohol sterilized steel blade from sample surfaces, following previously agreed ageing periods.

For alcoholic extraction methanol: water (volume ratio 4:1) (with methanol from REANAL, Budapest, Hungary) mixture was used and BRANSON 3510 ultrasonic bath (EMERSON, St. Louis, MO, USA) was applied for 20 min, after 5 mL extracting agent was mixed with the previously collected wood material (~0.05 g). The extract produced in this manner was centrifuged with a MiniSpin spinner (EPPENDORF, Hamburg, Germany) for 10 min (13,400 rpm).

The total phenolic content (TPC) was calculated based on the Folin–Ciocalteu method [53]. First, Folin–Ciocalteu reagent (VWR International, Debrecen, Hungary) (2.5 mL, tenfold dilution) was mixed with the extract (0.5 mL), and after 1 min application time, 2 mL (with concentration 0.7 M) Na_2CO_3 (VWR International, Debrecen, Hungary) solution was blended with the mixture. The reaction mixture was warmed in Memmert WNB 200 water bath (MEMMERT GmbH, Buechenbach, Germany) at 50 °C for 5 min. After warming in water bath, the solutions were cooled in cold water bath, until the temperature of solutions reached ~25 °C. For determination of total phenol content Metertech SP 8001 spectrophotometer (METERTECH Inc., Taipei, China) was used at 760 nm wavelength and as standard, quercetin (SIGMA-ALDRICH, St. Louis, MO, USA) was chosen.

Total soluble carbohydrate content (TSCC) was calculated based on method of Dubois et al. [54]. Firstly, phenol solution (REANAL, Budapest, Hungary) (0.5 mL; dilution: 5%) was mixed to the extract (0.5 mL). After mixing to the solution 2.5 mL concentrated sulfuric acid (REANAL, Budapest, Hungary), sealed test tubes were hold for 10 min at room temperature, followed by a second cooling for 20 min, in a 25 °C temperature water bath. Total soluble carbohydrate content was measured by using Metertech SP 8001 spectrophotometer (METERTECH Inc., Taipei, China) at 490 nm. During the total soluble carbohydrate content measurements, glucose was used as standard (SIGMA-ALDRICH, St. Louis, MO, USA). For the determination of TPC and TSCC, 3-3 replicates were analyzed.

2.5. Statistical Analysis

Statistical analysis was performed in Table 1, namely the analysis of variance (ANOVA) and t-test in order to explore significant variations between the following groups (measured data). No real significant differences between groups were considered in the ANOVA test when the value of statistic was close to 1 and relative low statistic value in case of t-test.

Table 1. The statistical analysis on each measured parameter.

Beech, Spruce, Birch, Sessile oak	†DWCA-TFC DWCA-TSCC ‡DMCA-TFC DMCA-TSCC TFC-TSCC DWCA-DMCA

†DWCA: Distilled water contact angle, ‡DMCA: Diiodomethane contact angle.

Datasets are given in publicly available database.

3. Results

3.1. Contact Angle Changes due to Artificial Ageing

The changes occurring in the development of the contact angles measured with different test liquids are different (bi-distilled water and diiodomethane) during ageing. The character of the development of contact angles is similar on the different wood species (Figure 2).

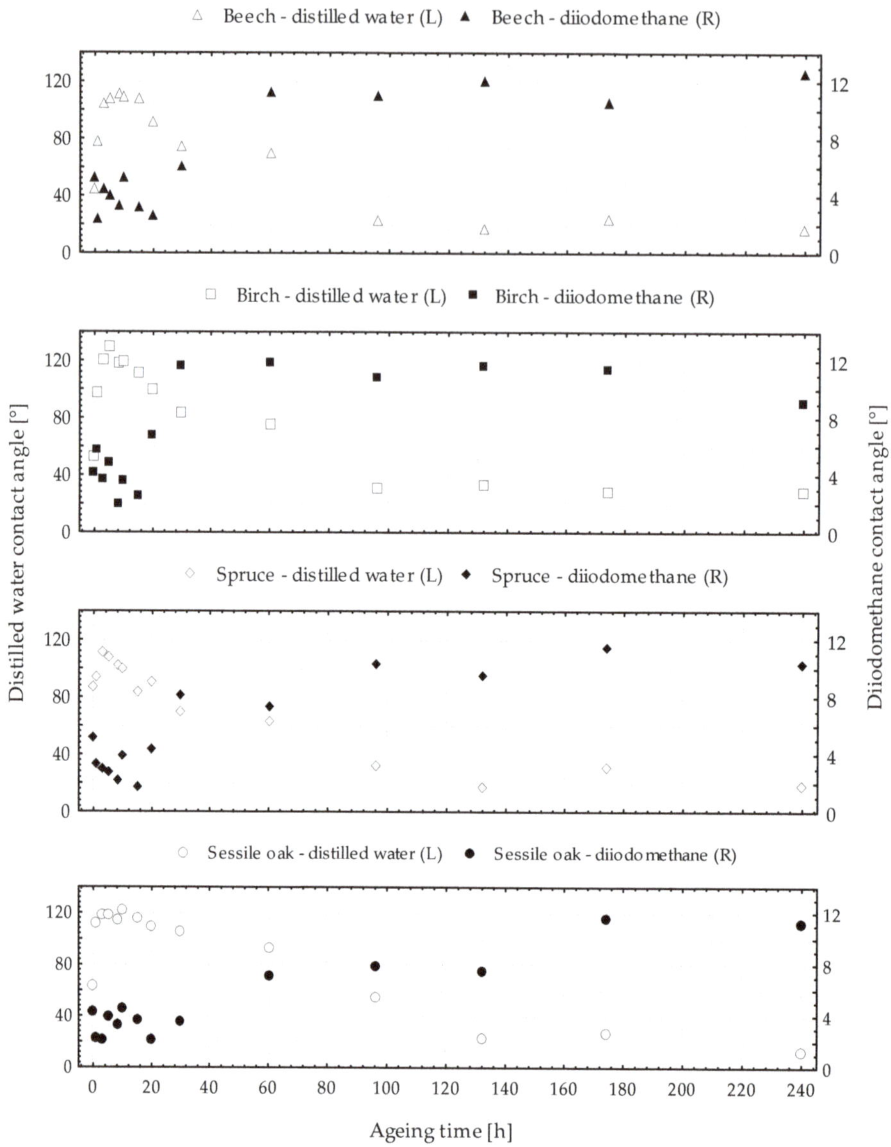

Figure 2. Figure 2. Contact angle values of distilled water and diiodomethane during artificial ageing, L—left y-axis, R—right y-axis.

There is some difference within the control values (0h) of distilled water contact angles measured on different wood species (Figure 2 and Table 2). Higher contact angle values were detected on sessile oak material, associated with higher extractive content, according to previous studies [15–17]. Since it has significant influence on liquid contact angle results, moisture content determinations were performed in order to monitor wood surface layer MC changes due to artificial ageing. MC results clearly show that, after the first three hours of artificial radiation, there was no difference between two consecutive measurements, in the case of all wood species. The results clearly show the process of changes during the total duration of artificial ageing, which takes place similarly in the case of all wood species: contact angle of distilled water increases in the first 8 h. Consequently, until the end of ageing (240 h), it has a slightly decreasing tendency.

Table 2. The average values of contact angle and its spread.

Ageing Time [h]	Distilled Water Contact Angle Values [°]							
	Beech		Spruce		Birch		Sessile oak	
	$\bar{x}$	SD	$\bar{x}$	SD	$\bar{x}$	SD	$\bar{x}$	SD
0	45.30	3.86	87.05	5.33	52.91	3.19	63.30	3.82
1	78.61	3.12	93.69	5.07	97.84	9.24	112.47	2.52
3	104.95	3.42	111.83	5.54	120.22	4.01	118.81	3.43
5	107.97	2.68	107.83	3.10	129.43	3.91	118.57	1.61
8	111.92	3.08	101.96	2.78	118.83	4.32	114.48	3.18
10	109.05	3.21	100.39	3.08	119.92	4.31	122.64	2.30
15	108.02	2.59	83.55	6.45	111.37	10.55	115.53	3.05
20	91.98	3.55	90.23	4.15	100.21	5.47	109.54	3.43
30	74.15	5.21	70.41	5.27	84.34	2.34	105.79	2.93
60	70.11	3.19	63.00	2.63	76.09	2.91	93.49	7.55
96	23.58	4.29	32.33	5.63	31.00	4.98	54.97	8.66
132	17.72	1.39	17.42	1.68	32.87	3.32	23.15	2.72
174	24.46	3.94	31.34	3.92	28.79	3.95	26.48	3.09
240	17.09	1.68	19.03	2.20	28.48	3.58	13.44	2.79
Ageing time [h]	Diiodomethane Contact Angle Values [°]							
	Beech		Spruce		Birch		Sessile oak	
	$\bar{x}$	SD	$\bar{x}$	SD	$\bar{x}$	SD	$\bar{x}$	SD
0	5.24	1.11	5.23	0.80	4.19	0.44	4.34	0.75
1	2.36	0.54	3.30	0.69	5.80	1.37	2.35	0.68
3	4.54	1.03	2.97	0.95	3.68	1.03	2.13	1.00
5	3.98	1.14	2.81	0.63	4.79	0.87	4.00	0.63
8	3.39	0.68	2.19	0.63	1.97	0.71	3.39	0.63
10	5.28	0.71	3.94	0.84	3.51	0.70	4.57	0.65
15	3.24	0.76	1.79	0.75	2.48	0.70	3.76	0.74
20	2.70	0.69	4.34	1.66	6.75	1.33	2.11	0.56
30	6.12	1.17	8.19	0.71	11.62	1.33	3.59	1.36
60	11.28	1.51	7.38	1.25	11.85	1.58	7.15	0.52
96	10.99	1.51	10.31	1.32	10.77	1.11	7.96	0.58
132	12.01	1.33	9.59	1.03	11.63	1.11	7.50	0.61
174	10.52	1.99	11.51	1.72	11.33	1.86	11.63	0.97
240	12.60	1.54	10.33	1.54	9.04	1.58	11.22	1.62

According to control values (0 h) and finally measured (after 240 h artificial radiation) distilled water contact angle values of different wood species can be concluded that the control values (0 h) are higher in case of all four wood species. Those results lead to the conclusion that higher wettability is characteristic to the wood surfaces at the end of artificial ageing (Figure 2). In addition, evaluating the total duration of artificial ageing, contact angle of both polar and disperse distilled water changes to a greater extent than the contact angle of the solely disperse diiodomethane.

3.2. *Total Phenolic and Total Soluble Carbohydrate Content of Artificially Aged Wood Surfaces*

Total phenolic content examinations were performed to detect quantity changes of phenolic extractive compounds of different wood materials (Figure 3). Phenolic compounds significantly influence the wetting of polar and disperse liquid drops, together with the wettability of wood surfaces.

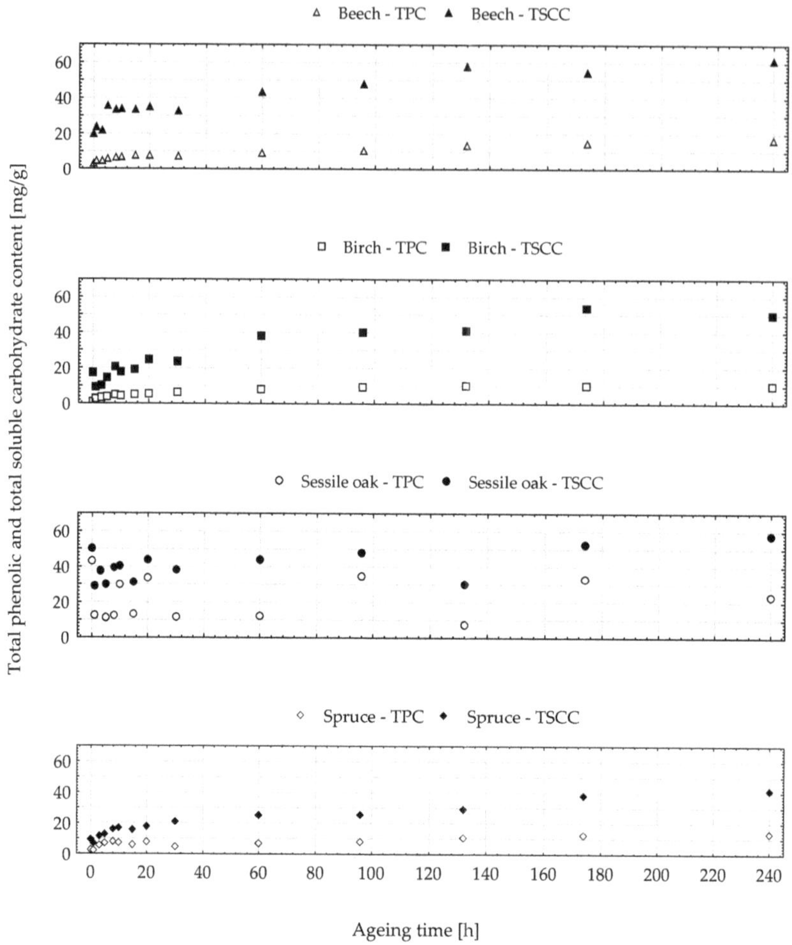

Figure 3. Total phenolic and total soluble carbohydrate contents as function of artificial ageing.

The results of total phenolic and total soluble carbohydrate content examinations show major difference in their measure. Total phenolic content of different wood species, except sessile oak, increases from a value about 1 mg/g to about 15 mg/g, regarding the total 240 h of artificial radiation. Parallel with the changes of total phenolic content, the total soluble carbohydrate content (for all four wood species) increases from a value of 7 mg/g to about 45–60 mg/g. The results clearly show that the carbohydrate content changes in 3–4 times higher measure than the phenolic compounds (Table 3).

Table 3. The average values of TPC and TSCC and its spread.

Ageing Time [h]	TFC Values [mg/g]							
	Beech		Spruce		Birch		Sessile oak	
	x̄	SD	x̄	SD	x̄	SD	x̄	SD
0	2.923	0.042	2.519	0.096	0.729	0.083	43.070	2.842
1	4.706	0.085	2.127	0.092	2.533	0.101	12.481	1.517
3	4.691	0.042	5.479	0.155	3.325	0.095	37.380	1.187
5	5.974	0.178	6.805	0.257	3.665	0.066	11.201	0.090
8	6.702	0.055	7.780	0.205	4.910	0.183	12.311	0.466
10	6.867	0.191	7.119	0.227	4.328	0.117	29.842	1.669
15	7.770	0.072	5.775	0.253	5.017	0.136	13.215	1.001
20	7.819	0.113	7.732	0.175	5.328	0.199	33.759	1.891
30	7.295	0.139	4.683	1.907	6.268	0.330	11.653	1.045
60	9.113	0.123	6.807	0.439	8.108	0.258	12.208	0.413
96	10.615	0.136	8.003	0.478	9.309	0.261	34.748	0.870
132	13.613	0.196	10.804	0.244	10.095	0.114	7.503	0.239
174	14.963	0.370	12.418	0.886	10.208	0.254	33.061	0.905
240	17.041	0.199	13.371	0.692	10.281	0.401	23.380	0.495
	TSCC Values [mg/g]							
Ageing Time [h]	**Beech**		**Spruce**		**Birch**		**Sessile oak**	
	x̄	SD	x̄	SD	x̄	SD	x̄	SD
0	19.762	0.504	9.255	0.402	17.177	2.157	50.172	3.595
1	23.731	4.575	6.773	0.755	9.078	0.829	28.889	0.976
3	21.906	2.880	11.468	1.086	10.058	2.493	37.681	7.343
5	35.616	5.721	12.569	2.080	14.373	2.557	29.903	3.324
8	33.638	2.732	15.849	2.621	20.624	1.430	39.438	10.706
10	33.858	2.186	16.681	2.880	17.628	0.174	40.276	0.554
15	33.459	3.573	15.555	3.330	19.073	0.561	31.259	2.626
20	34.883	0.163	17.769	3.698	24.459	0.950	43.858	2.782
30	32.876	5.870	20.903	3.423	23.505	3.657	38.251	5.158
60	43.517	3.090	25.110	2.013	37.897	2.891	44.040	0.488
96	47.914	1.348	25.370	7.367	39.954	0.797	47.991	1.972
132	58.020	3.653	29.226	4.664	41.100	3.795	30.234	1.348
174	54.711	1.643	38.010	1.481	53.740	5.197	52.610	6.920
240	61.471	6.214	41.353	8.009	49.826	2.215	57.729	3.781

4. Discussion

The quantitative change described above is consistent regarding the change of the measured contact angle values. The measured contact angle, TPC and TSCC values in case of different wood species show similar trend under the entire artificial ageing period.

Despite the fact that the total phenol content increases almost 15-times during 240 h of artificial xenon radiation and the total soluble carbohydrate content increases only 6–8 times, the substantially higher presence of carbohydrate has a greater influence on the contact angle of the liquid drops. Based on the decrease of contact angles an improved wood surface wettability can be concluded.

The most reactive wood components during photodegradation are the lignin and the extractives [39,41,42]. The reason for contact angle changes in the first 15–20 h of artificial radiation is that the decomposition of lignin and extractives begins, and the total phenolic content increases. In the subsequent radiation interval begins the decay of cellulose and hemicelluloses, which are less sensitive to photodegradation. The total soluble carbohydrate content increases compensate the effect of hydrophobic phenolic compounds and this is manifest in the diminution of strong variation in the contact angle values. As cellulose and hemicelluloses are present on the wood surface (and in the whole wood material) in essentially larger quantities, their hydrophilic nature is strongly manifested in the next radiation interval proved, also by decreasing contact angle values. The observed chemical changes were further investigated with Fourier-transform infrared spectroscopy and these results will be published in the near future.

Overall, the performed t-test statistical analysis resulted in values close to 0 in each combination, which indicate that similarity exists between the measured data sets, and supports the observation that the contact angle value, for example, is susceptible to follow surface chemical changes. The ANOVA test also indicates that there is a real difference between the tested, measured data groups in each combination, as the value was close to zero.

5. Conclusions

Chemical changes of the wood surface layer, which can be well monitored by using total phenolic and soluble carbohydrate content examinations, are influenced by the photodegradation of the main wood components: cellulose, hemicelluloses, and lignin. According to test results of TPC and TSCC experiments, by using the 75-μm thin layer of wood specimens, both phenolic and carbohydrate content alter to a great extent due to photodegradation. The quantity increases of total soluble carbohydrates, which gives information about the degradation of cellulose and hemicelluloses, and their basically greater presence in wood material, influences the contact angle of liquid drops more than the total phenolic compounds.

To our best knowledge, there is no article that reports on TPC and TSCC evaluation considering the surface detached wood particles observed under artificial ageing. The present study shows that the preparation method and evaluation of TPC and TSCC can be significant. This could be a great advantage and novel technique, thus representing a good contribution to the surface science of wood. In that sense, our attempt to draw attention to the method used might also be useful to researchers from the other fields of the wood science, like surface treatment and the gluing of wood.

Author Contributions: Conceptualization, E.A.P. and L.C.; methodology, T.H. and L.C.; software, E.A.P. and A.N.M; validation, E.A.P., A.N.M. and T.H.; formal analysis, C.C.; investigation, E.A.P. and A.N.M.; resources, C.C.; data curation, L.C.; writing—original draft preparation, E.A.P.; writing—review and editing, E.A.P. and L.C.; visualization, E.A.P. and A.N.M.; supervision, C.C.; project administration, C.C.; funding acquisition, L.C. All authors have read and agreed to the published version of the manuscript.

Funding: This research was supported by the European Union and the State of Hungary, co-financed by the European Social Fund in the framework of TAMOP 4.2.4. A/2-11-1-2012-0001 National Excellence Program.

Acknowledgments: This article was made in frame of "the EFOP-3.6.1-16-2016-00018—Improving the role of research + development + innovation in the higher education through institutional developments assisting intelligent specialization in Sopron and Szombathely". IKEA Industry Sopron Ltd. for wood material supply and the carpenter workshop team of US-SKF for their skillful work are gratefully acknowledged.

Conflicts of Interest: The authors declare no conflict of interest.

References

1. Diouf, P.-N.; Merlin, A.; Perrin, D. Antioxidant properties of wood extracts and colour stability of woods. *Ann. For. Sci.* **2006**, *63*, 525–534. [CrossRef]
2. Rowell, R.M.; Pettersen, R.; Han, J.S.; Rowell, J.S.; Tshabalala, M.A. Cell wall chemistry. In *Handbook of Wood Chemistry and Wood Composites Part I*, 1st ed.; Rowell, R.M., Ed.; Taylor and Francis CRC Press: Boca Raton, FL, USA, 2005; Volume 3, p. 53.
3. Syofuna, A.; Banana, A.Y.; Nakabonge, G. Efficiency of natural wood extractives as wood preservatives against termite attack. *Maderas Sci. Technol.* **2012**, *14*, 155–163. [CrossRef]
4. Umezawa, T. Chemistry of extractives. In *Wood and Cellulosic Chemistry*, 2nd ed.; Hon, D.N.S., Shiraishi, N., Eds.; Marcell-Decker: New York, NY, USA, 2001; pp. 213–241.
5. Aloui, F.; Ayadi, N.; Charrier, F.; Charrier, B. Durability of European oak (*Quercus petraea* and *Quercus robur*) against white rot fungi (*Coriolus versicolor*): Relations with phenol extractives. *Holz Als Roh Und Werkst.* **2004**, *62*, 286–290. [CrossRef]
6. Trapp, S.; Miglioranza, K.S.; Mosback, H. Sorption of lipophilic organic compounds to wood and implications for their environmental fate. *Environ. Sci. Technol.* **2001**, *35*, 1561–1566. [CrossRef]

7. Young, R.A. Wettability of wood pulp fibers. *Wood Fiber Sci.* **1976**, *8*, 120–128.
8. Hse, C.; Kuo, M. Influence of extractives on wood gluing and finishing—A review. *For. Prod. J.* **1988**, *38*, 52–56.
9. Kajita, H.; Skaar, C. Wettability of the surfaces of some American softwoods species. *Mokuzai Gakkaishi* **1992**, *38*, 516–521.
10. Wålinder, M.; Gardner, D.J. Acid-base characterization of wood and selected thermoplastics. *J. Adhes. Sci. Technol.* **2002**, *16*, 1625–1649. [CrossRef]
11. Visi-Rajczi, E. The Production, Accumulation and Distribution of Beech (Fagus sylvatica L.) Extractives (in Hungarian). Ph.D. Thesis, University of West-Hungary, Sopron, Hungary, 2008.
12. River, B.H.; Vick, C.B.; Gillespie, R.H. Treatise on adhesion and adhesives. In *Wood as an Adherend*, 1st ed.; Minford, J.D., Ed.; Marcel Decker: New York, NY, USA, 1991; Volume 7, pp. 134–162.
13. Nussbaum, R.M. Surface Interactions of Wood with Adhesives and Coatings. Ph.D. Thesis, KTH—Royal Institute of Technology, Stockholm, Sweden, 2001.
14. Nguyen, T.; Johns, W.E. The effects of aging and extractives on the surface free energy of Douglas-fir and redwood. *Wood Sci. Technol.* **1979**, *13*, 29–40. [CrossRef]
15. Jaić, M.; Živanović, R.; Stevanović-Janežić, T.; Dekanski, A. Comparison of surface properties of beech and oak wood as determined by ESCA method. *Holz Als Roh Und Werkst.* **1996**, *54*, 37–41. [CrossRef]
16. Mantanis, G.I.; Young, R.A. Wetting of wood. *Wood Sci. Technol.* **1997**, *31*, 339–353. [CrossRef]
17. Gindl, M.; Sinn, G.; Gindl, W.; Reiterer, A.; Tschegg, S. A comparison of different methods to calculate the surface free energy of wood using contact angle measurements. *Coll. Surf.* **2001**, *181*, 279–287. [CrossRef]
18. White, M.S.; Ifju, G.; Johnson, J.A. The role of extractives in the hydrophobic behavior of loblolly pine rhytidome. *Wood Fiber Sci.* **1974**, *5*, 353–363.
19. Rossi, D.; Rossi, S.; Morin, H.; Bettero, A. Within-tree variations in the surface free energy of wood assessed by contact angle analysis. *Wood Sci. Technol.* **2012**, *46*, 287–298. [CrossRef]
20. Nguyen, T.; Johns, W.E. Polar and dispersion force contributions to the total surface free energy of wood. *Wood Sci. Technol.* **1979**, *12*, 63–74. [CrossRef]
21. Kalnins, M.A.; Katzenberger, C.; Schmieding, S.A.; Brooks, J.K. Contact angle measurement on wood using videotape technique. *J. Colloid Interface Sci.* **1988**, *125*, 344–346. [CrossRef]
22. Kishino, M.; Nakano, T. Artificial weathering of tropical woods. Part 1: Changes in wettability. *Holzforschung* **2004**, *58*, 552–557. [CrossRef]
23. Amorim, M.R.S.; Ribeiro, P.G.; Martins, S.A.; del Menezzi, C.H.S.; de Souza, M.R. Surface wettability and roughness of 11 Amazonian tropical hardwoods. *Florest. Ambient.* **2013**, *20*, 99–109. [CrossRef]
24. Subedi, D.P. Contact angle measurement for the surface characterization of solids. *Himal. Phys.* **2011**, *2*, 1–4. [CrossRef]
25. Snoeijer, J.H.; Andreotti, B. A microscopic view on contact angle selection. *Phys. Fluid.* **2008**, *20*, 1–11. [CrossRef]
26. Rubina, T.; Stalidzans, E.; Diminš, F.; Kuka, P.; Morozovs, A. Determination of the characteristic value of the contact angle and surface energy for wood. *Latv. Lauksaimn. Univ. Raksti* **2009**, *22*, 100–112. [CrossRef]
27. Good, R.J.; Chaudhury, M.K.; Yeung, C. A new approach for determining roughness by means of contact angles on solids. In *First International Congress on Adhesion Science and Technology*; Ooij, W.J., Anderson, H.R., Eds.; Mit. Festschr.: Utrecht, The Netherlands, 1998; pp. 181–197.
28. Wålinder, M.; Ström, G. Measurement of wood wettability by the wilhelmy method—Part 2. determination of apparent contact angles. *Holzforschung* **2001**, *55*, 33–41. [CrossRef]
29. Jirouš-Rajković, V.; Turkulin, H.; Miller, E.R. Depth profile of UV-induced wood surface degradation. *Surf. Coat. Int. Part B Coat. Trans.* **2004**, *87*, 235–308. [CrossRef]
30. Gindl, M.; Reiterer, A.; Sinn, G.; Stanzl-Tschegg, S.E. Effects of surface ageing on wettability, surface chemistry and adhesion of wood. *Holz Als Roh Und Werkst.* **2004**, *62*, 273–280. [CrossRef]
31. Herczeg, A. The wettability of wood. *For. Prod. J.* **1965**, *15*, 499–505.
32. Nussbaum, R.M. Natural surface inactivation of scots pine and norway spruce evaluated by contact angle measurements. *Holz Als Roh Und Werkst.* **1999**, *57*, 419–424. [CrossRef]
33. Borgin, K.; Faix, O.; Schweers, W. The effect of aging on lignins of wood. *Wood Sci. Technol.* **1975**, *9*, 207–211. [CrossRef]

34. Wolkenhauer, H.; Avramidis, G.; Hauswald, E.; Militz, H.; Viol, W. Sanding vs. plasma treatment of aged wood—A comparison with respect to surface energy. *Int. J. Adhes. Adhes.* **2009**, *29*, 18–22. [CrossRef]
35. Wiedenhoeft, A.C.; Miller, R.B. Structure and function of wood. In *Handbook of Wood Chemistry and Wood Composites Part I*, 1st ed.; Rowell, R.M., Ed.; Taylor and Francis CRC Press: Boca Raton, FL, USA, 2005; Volume 2, pp. 29–38.
36. Campbell, M.M.; Sederoff, R.R. Variation in lignin content and composition. *Plant Physiol.* **1996**, *110*, 3–13. [CrossRef]
37. Koch, G.W.; Sillett, S.C.; Jennings, G.M.; Davis, S.D. The limits to tree height. *Nature* **2004**, *428*, 851–854. [CrossRef]
38. Williams, R.S.; Jourdain, C.; Daisey, G.I.; Springate, R.W. Wood properties affecting finish service life. *J. Coat. Technol.* **2000**, *72*, 35–42. [CrossRef]
39. Müller, U.; Rätzsch, M.; Schwanninger, M.; Steiner, M.; Zöbl, H. Yellowing and IR-changes of spruce wood as result of UV-irradiation. *J. Photochem. Photobiol. B Biol.* **2003**, *69*, 97–105. [CrossRef]
40. Németh, K. *Degradation of Wood (in Hungarian)*, 1st ed.; Mezőgazdasági Szaktudás Kiadó: Budapest, Hungary, 1998; pp. 53–58.
41. Pandey, K.K. A note on the influence of extractives on the photo-discoloration and photo-degradation of wood. *Polym. Degrad. Stab.* **2005**, *87*, 375–379. [CrossRef]
42. Pandey, K.K. Study of the effect of photo-irradiation on the surface chemistry of wood. *Polym. Degrad. Stab.* **2005**, *90*, 9–20. [CrossRef]
43. Tolvaj, L.; Varga, D. Photodegradation of timber of three hardwood species caused by different light sources. *Acta Silv. Lignaria Hung.* **2012**, *8*, 145–155. [CrossRef]
44. Ramirez-Ramirez, G.; Lubbers, S.; Voilley, A.; Charpentier, C.; Feuillat, M.; Chassagne, D. Aroma compound sorption by oak wood in a model wine. *J. Agric. Food Chem.* **2001**, *49*, 3893–3897. [CrossRef]
45. Salame, I.I.; Bandosz, J.T. Role of surface chemistry in adsorption of phenol on activated carbons. *J. Colloid Interface Sci.* **2003**, *264*, 307–312. [CrossRef]
46. Ayadi, N. Vieillissement climatique d'un système bois-vernis-absorbeur UV inorganique (in French with English abstract), Université de Nantes—Sciences des Matériaux, France. 2004. Available online: https://scholar.google.com.hk/scholar?hl=zh-CN&as_sdt=0%2C5&q=Vieillissement+climatique+d%E2%80%99un+syst%C3%A8me+bois-vernis-absorbeur+UV+inorganique+%28in+French+with+English+abstract%29%2C+Universit%C3%A9+de+Nantes%E2%80%94Sciences+des+Mat%C3%A9riaux%2C+France%2C+2004&btnG=#d=gs_cit&u=%2Fscholar%3Fq%3Dinfo%3AzsCJnj1XHH4J%3Ascholar.google.com%2F%26output%3Dcite%26scirp%3D0%26hl%3Dzh-CN (accessed on 9 March 2020).
47. Anjos, J.P.; Cardoso, M.G.; Saczk, A.A.; Dórea, H.S.; Santiago, W.D.; Machado, A.M.R.; Zacaroni, L.M.; Nelson, D.L. Evolution of the concentration of phenolic compounds in cachaça during aging in an oak (Quercus sp.) barrel. *J. Braz. Chem. Soc.* **2011**, *22*, 1307–1314. [CrossRef]
48. Gindl, M.; Sinn, G.; Stanzl-Tschegg, S. The effects of ultraviolet light exposure on the wetting properties of wood. *J. Adhes. Sci. Technol.* **2006**, *20*, 817–828. [CrossRef]
49. Kuo, M.; Hu, N. Ultrastructural changes of photodegradation of wood surfaces exposed to UV. *Holzforschung* **1991**, *45*, 347–353. [CrossRef]
50. Williams, R.S. Weathering of wood. In *Handbook of Wood Composites*, 1st ed.; Rowell, R.M., Ed.; Taylor and Francis CRC Press: Boca Raton, FL, USA, 2005; Volume 7, pp. 142–178.
51. Hon, D.N.S.; Ifju, G. Measuring penetration of light into wood by detection of photoinduced free radicals. *Wood Sci.* **1978**, *11*, 118–127.
52. Németh, K.; Faix, O. Beobachtung der photodegradation des Holzes durch quantitative drift-Spektroskopie. *Holz Als Roh Und Werkst.* **1994**, *52*, 261–266. [CrossRef]

53. Singleton, V.L.; Rossi, J.A. Colorimetry of total phenolics with phosphomolybdic-phosphotungstic acid reagents. *Am. J. Enol. Vitic.* **1965**, *16*, 144–158.
54. Dubois, M.; Gilles, K.A.; Hamilton, J.K.; Rebers, P.A.; Smith, F. Colorimetric method for determination of sugars and related substances. *Anal. Chem.* **1956**, *28*, 350–356. [CrossRef]

coatings

MDPI

Article

Shelling of Growth Rings at Softwood Surfaces Exposed to Natural Weathering

Lukie H. Leung [1] **and Philip D. Evans** [1,2,*]

[1] Department of Wood Science, Faculty of Forestry, University of British Columbia, Vancouver, BC V6T 1Z4, Canada; mrlukieleung@gmail.com

[2] Department of Applied Mathematics, The Australian National University, Canberra, ACT 0200, Australia

* Correspondence: phil.evans@ubc.ca; Tel.: +1-604-822-0517; Fax: +1-604-822-9159

Received: 13 August 2020; Accepted: 2 September 2020; Published: 5 September 2020

Abstract: Shelling is the delamination of growth rings and the projection of woody tissue from wood surfaces. Shelling disrupts coatings and makes refinishing difficult, and a better understanding of the phenomenon is needed to help alleviate its unwanted effects. We tested whether confocal profilometry could quantify shelling in flat-faced and profiled-faced western larch deckboards exposed to natural weathering and examined the effects of growth-ring orientation and angle on shelling. Confocal profilometry was able to quantify shelling in both deckboard types. Shelling developed at the surface of flat-faced deckboards oriented pith-side-up, whereas it was absent from boards oriented bark-side-up. We found an inverse correlation between the height of shelled growth rings and the angle of growth rings to the surface of flat-faced boards. Shelling occurred in profiled-faced boards oriented pith-side-up due to the delamination of growth ring tips and projection of latewood from wood surfaces. A superficially similar although less pronounced phenomenon was seen in profiled-faced boards oriented bark-side-up. The shelling of profiled-faced boards oriented pith-side-up created lanceolate-shaped slivers of latewood that projected from the peaks of profiles. Some of these latewood tips were sharp and, for this reason, we suggest that profiled-faced western larch deckboards should always be oriented bark-side-up rather than pith-side-up.

Keywords: shelling; western larch; confocal profilometry; profiling; growth rings; latewood; earlywood; pith-side-up; bark-side-up; natural weathering

1. Introduction

Wood that is exposed outdoors is subject to surface degradation by solar radiation (U.V. and visible light), water, heat, environmental pollutants, and mould [1]. The most obvious manifestation of such degradation is the greying of wood exposed outdoors [2]. Another highly visible effect of "weathering" is the surface cracking (checking) of wood [3]. Checks develop when moisture-induced surface stresses exceed the tensile strength of wood perpendicular to the grain [4]. The severity of checking depends, in part, on the orientation of growth rings to wood surfaces exposed to the weather. For example, checking is less severe in coniferous wood (softwood) whose growth ring are perpendicular to the exposed surface (radial or quarter-sawn boards) when compared to wood whose growth rings are mainly parallel to the surface (tangential or flat-sawn boards) [5]. Checks in quarter-sawn softwoods often develop at the interface between denser bands of summer wood (latewood) and lower density springwood (earlywood) [6]. In flat-sawn softwoods, checks develop in and propagate radially into wood via rays [4]. Flat-sawn wooden boards can be classified into ones where growth rings either curve away from, or curve towards, their upper wide surface [7] (Figure 1). These boards are commonly known as bark-side-up and pith-side-up boards, respectively [7] (Figure 1). The checking of wood during exposure to the weather is less pronounced in wooden boards that are oriented pith-side-up [5,8].

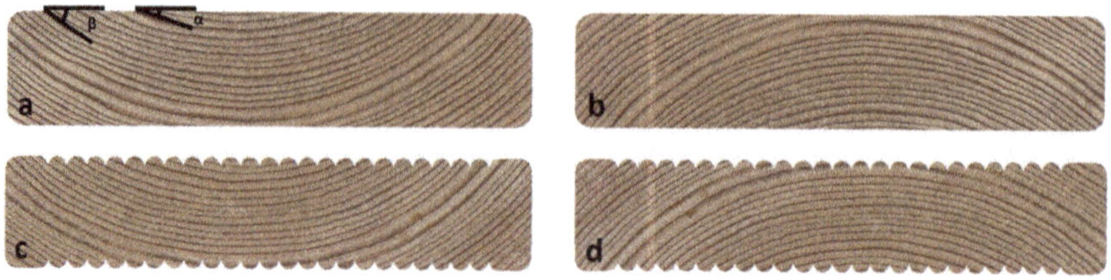

Figure 1. Orientation of growth rings to the upper exposed surface of flat-sawn (tangential) western larch deckboards. The darker bands within growth rings are latewood and the lighter bands are earlywood: (**a**) flat board with growth rings that curve towards the exposed upper surface. These are known as pith-side-up boards. Growth rings can make shallow (α) or high (β) angles to the exposed upper surface; (**b**) flat board with growth rings that curve away from the exposed upper surface. These are known as bark-side-up boards; (**c**) profiled pith-side-up board; and, (**d**) profiled bark-side-up board.

In addition to checking, another less commonly known defect, shelling or loosened grain, develops at wood surfaces, especially those that are exposed to the weather [7,9–11]. Shelling occurs when mechanical or moisture-induced stresses at the surface of flat-sawn boards are sufficiently large to cause delamination of growth rings at the boundary between latewood and earlywood [9,10]. Such delamination of growth rings results in the projection of latewood above the surface of adjacent earlywood. Two forms of shelling have been observed at flat-sawn (tangential) surfaces that result from either the separation of tips or edges of latewood from earlywood, respectively [9,10]. Shelling mainly develops at the surface of boards oriented pith-side-up. It can also occur beneath opaque coatings and disrupt paint films on siding (cladding), and this is one of the reasons why wooden siding is fixed to building with growth rings oriented bark-side facing outwards [10,11]. Shelling, particularly shelled tips of latewood, can create sharp splinters that protrude from the surface of wood. These splinters are objectionable and dangerous according to Koehler [9], particularly when they occur at the surface of deckboards that people walk on. Shelling also makes refinishing and cleaning of wood surfaces difficult [9,10]. Shelling is most pronounced in softwoods with a pronounced difference in cell wall thickness between latewood and earlywood, for example, species, such as southern pine (*Pinus* spp.), larch (*Larix* spp.), and Douglas fir (*Pseudotsuga menziesii* (Mirb.) Franco) [9,10,12], but it has also been observed in hardwood with relatively uniform structure, such as yellow poplar (*Liriodendron tulipfera* L.) [10]. These wood species are very important commercially. Furthermore, the wood products that are negatively affected by shelling such as decking and siding are especially important. For example, the market for wooden decking and siding in the USA had a combined value in 2018 of ~US 2.8 billion, and each year over 300 million square feet (~28 million m^2) of wood siding is installed on houses in the USA [13,14]. However, wooden siding and decking are losing market share, because they require more maintenance particularly refinishing and cleaning than products made from alternative materials, such as plastic and composites (wood-plastic and wood-fibre cement composites) [13,14]. Hence, there is a need to better understand and develop solutions to problems such as shelling that reduce coating durability and increase the frequency of refinishing of wood products used for siding and decking.

Our understanding of shelling is based on a small number of studies [7,9,10], but, unlike grain raising, which was until recently a similarly neglected surface phenomenon that affects the performance of coatings on wood, no method has been developed to quantify shelling at weathered wood surfaces. Grain raising was first quantified using stylus profilometry [15], and recent studies have used non-contact profilometry to quantify and image grain raising [16,17], and examine the shape recovery of earlywood and latewood beneath machined and painted radiata pine (*Pinus radiata* D. Don) panels soaked in water [18,19]. The latter research suggests that the current generation of non-contact profilometers will be able to quantify shelling at weathered wood surfaces and provide insights into its origins. We test this hypothesis in this paper. We use confocal profilometry and macro-photography to quantify and image shelling of flat-faced western larch (*Larix occidentalis* Nutt.) deckboards (hereafter called flat deckboards) exposed to natural weathering for seven months. Profilometry and scanning electron microscopy were also used to image shelling of flat alkaline copper quaternary (ACQ) treated Douglas fir, western hemlock (*Tsuga heterophylla* (Raf.) Sarg.), and white spruce (*Picea glauca* (Moench) Voss) deckboards exposed to the weather for 18 months. We extend our observations of shelling of flat deckboards to include profiled-faced larch deckboards (hereafter called profiled deckboards) that were exposed to the weather because there are no reports in the literature on the shelling of profiled deckboards, even though profiling is widely used in Australia, Europe, Japan, and New Zealand to reduce the negative effects of checking on the appearance of deckboards exposed outdoors [20,21].

2. Materials and Methods

2.1. Preparation and Weathering of Larch Deckboards

Six flat-sawn, J-grade, kiln-dried, and dressed western larch boards cut from trees growing in the interior of British Columbia, BC, Canada, and measuring 38 × 140 × 3658 mm^3, were donated by Tolko's Lavington Planer Mill Ltd. (6200 Jeffers Drive, Lavington, British Columbia, BC, Canada). Each parent board was cross-cut using a pendulum saw (Stromab ps 50/f, Campagnola Emilia, Italy) to produce six samples, each 600 mm in length. Samples were then randomly assigned to one of five deckboard profiles or the flat control, Table 1. Sample boards were then profiled on both sides using a moulding machine (Weinig Powermat 700, Michael Weinig Inc., Mooresville, NC, USA) equipped with a 125 mm diameter, two-wing, cylindrical rotary cutter head (Great-Loc SG Positive Clamping Universal Tool System, Great Lakes Custom Tool Mfg. Inc., Pestigo, WI, USA). After double-sided profiling, each sample was cut into two using a pendulum saw, as above. The sub-samples were oriented either pith-side-up or bark-side-up during the subsequent natural weathering trial. A total of seventy-two deckboard samples each with a final dimension of 32 × 130 × 292 mm^3 were manufactured. The ends of deckboard samples were coated with an epoxy sealer (Intergard® 740, International Paint Singapore Pty Ltd., Jurong, Singapore) to prevent end checking of boards during conditioning and weathering. The decking samples were stored in a conditioning room at 20 ± 1 °C and 65 ± 5% r.h. for at least three months before the outdoor weathering trial.

Table 1. Geometry of profiles in western larch deckboards.

Profile Type [1]	Groove Depth (mm)	Groove Radius (mm)	Peak Radius (mm)
Flat	-	-	-
Rib	2.0	0.16	2.4
Short rib	1.5	0.16	2.4
Tall rib	2.5	0.15	2.2
Ribble	2.0	0.65	1.3
Ripple	2.0	1.0	1.2

[1] Number of peaks per 150 mm was 20 for all profiles except flat samples.

Profiled and flat deckboard sub-samples from the same parent western larch board (board 1, 2, 3, 4, 5, or 6) were screwed onto separate sub-frames made from preservative-treated 2 × 4-dimensional lumber. Each deckboard was spaced 10 mm apart from adjacent boards and fastened at its four corners onto the weathering rack using hidden galvanized screws (CAMO® fastening system, National Nail Corp. Grand Rapids, MI, USA, http://www.camofasteners.com/). A total of six mini-decks were constructed, each measuring 30 cm long, 177 cm wide, and 61 cm high. In addition, end boards made from preservative-treated lumber measuring ~40 × 90 × 300 mm^3 were screwed on to the two ends of each rack to prevent the edges of samples at the ends of racks from weathering. The test decks were then placed outdoors in Vancouver for seven months from March to October 2019. The deckboard samples were removed from racks and stored in a conditioning room at 20 ± 1 °C and 65 ± 5% r.h. for two weeks before profilometry measurements. After the weathering trial, samples measuring

$15 \times 15 \times 38$ mm^3 were cut from each of the six parent western larch boards. The basic density of these samples was calculated using their oven dry weight (obtained by oven drying samples at 105 °C to a constant weight) divided by their water-saturated volume (obtained by Archimedean displacement).

2.2. Measurement and Imaging of Shelling of Weathered Deckboards

The shelling of the flat western larch deckboard samples was characterized using a non-contact surface confocal profilometer equipped with a 3 mm probe (Altisurf 500®, Altimet, 298 Allée du Larry, 74200, Marin, France) [22]. The profilometer measured the heights of each deckboard sample along a 100 mm line at three locations; each line was offset approximately 16 mm from the edges of samples. Height data were extracted from line scans using the software ProfilmOnline (Filmetrics, KLA Co., San Diego, CA, USA). Heights of shelled growth rings were measured from the tip of the latewood projecting from the wood surface to the adjacent earlywood. In addition, a 625 mm^2 area on samples was scanned using the confocal profilometer and reconstructed using Altimet Premium (version 6.2) to provide three-dimensional (3D) images of growth rings. Profilometry was also used to image shelling of latewood tips occurring on the peaks of profiled western larch deckboard samples. The number of detached or raised latewood tips at the surface of profiled deckboard samples was counted, and the surfaces of flat and profiled boards were photographed using a Canon EOS 80D digital single lens reflex camera equipped with a Canon EF-S 35 mm f/2.8 macro lens. End-grain of flat deckboards with shelling were scanned using a desktop scanner (Hewlett-Packard Office Jet 6700, Palo Alto, CA, USA) at 600 dpi resolution, which was sufficient for subsequent image analysis. Scanned images were then examined with the image analysis software ImageJ (version 1.51q) (https://imagej.nih.gov/ij/) to measure growth ring angle to the surface of the corresponding shelled growth ring (Figure 1a). The number of growth rings per cm on the end grain of deckboard samples was measured using a transparent plastic ruler.

We also imaged shelling in flat Douglas fir, western hemlock, and white spruce deckboards that had been treated with a 1.8% alkaline copper quaternary (ACQ) preservative and exposed to natural weathering for 18 months. These boards were part of an experimental trial that examined effects of three factors, profile type, wood species (Douglas fir, western hemlock and white spruce) and growth ring orientation (boards with growth rings oriented either bark-side-up or pith-side-up) on the cupping and checking of deckboards during weathering. The materials and methods that were used to prepare and weather these deckboards are described in detail in a recent publication [8]. Flat weathered deckboards with growth rings oriented pith-side-up were selected, and shelled growth rings were imaged using confocal profilometry and macro-photography, as described above.

Scanning electron microscopy was used to examine the interface between latewood and earlywood in shelled growth rings in flat Douglas fir, western hemlock and white spruce deckboards. Samples measuring $15 \times 15 \times 15$ mm^3 and each containing a shelled growth ring were cut from deckboard samples using a small razor saw (Lee Valley Tools Ltd., Vancouver, BC, Canada, ultra-thin razor saw 60F0310). Samples were dried over silica gel at 20 ± 1 °C for 24 h and reduced in size to ~$5 \times 5 \times 8$ mm^3 using single-edged razor blades (gem surgical carbon steel blades, American Safety Razor Co. Cedar Knolls, NJ, USA) [23]. Samples with either transverse or tangential (weathered) surfaces facing uppermost were glued to separate aluminium stubs using Nylon nail polish as an adhesive. The stubs were sputter coated with an 8 nm layer of gold and they were then examined using a Zeiss Ultra Plus field emission scanning electron microscope (Carl Zeiss AG, Oberkochen, Germany) at an accelerating voltage of 5 kV, working distances of 14.8 to 15.9 mm, and a high vacuum (1.3×10^{-4} Pa). Secondary electron images of samples were obtained and saved as TIF files.

2.3. Experimental Design and Statistical Analysis of Data

Our experiment was designed to examine the effects of surface profiling and growth ring orientation on the checking and cupping of western larch deckboards that were exposed to natural weathering. The experiment was a randomized block design with two fixed factors: (a) Boards with different surface profiles, as defined and described previously, Table 1 [24], including three rib profiles (rib, short rib, and tall rib), and two wavy profiles (ribble and ripple): flat boards acted as a control; (b) growth ring orientation (concave, pith-side-up; and convex, bark-side-up). Six replicate blocks of twelve samples (6 profiles × 2 orientations × 6 replicates = 72 samples) were prepared and exposed to natural weathering. Linear regression was used to examine the relationship between growth ring angle and height (mm) of shelled growth rings in flat western larch boards oriented pith-side-up. Analysis of variance was used to examine the effects of growth ring orientation and profile type on numbers of shelled growth rings in profiled boards. The latter count data was transformed (square root) before final analysis. Statistical computation used the software, Genstat (v. 19). The results are presented graphically and Fisher's least significant difference (LSD) bar ($p < 0.05$) can be used to determine whether differences between individual means are statistically significant [25].

3. Results

3.1. Shelling of Growth Rings in Flat Western Larch Deckboards

Shelling was clearly evident at the surface of flat western larch deckboards exposed pith-side-up to the weather. The confocal profilometry height map and adjacent matching photographic image in Figure 2 show that shelling of western larch deckboards was due to oblique checking at the interface between latewood and earlywood and the projection of latewood above the deckboard surface, as in Figure 2a,b.

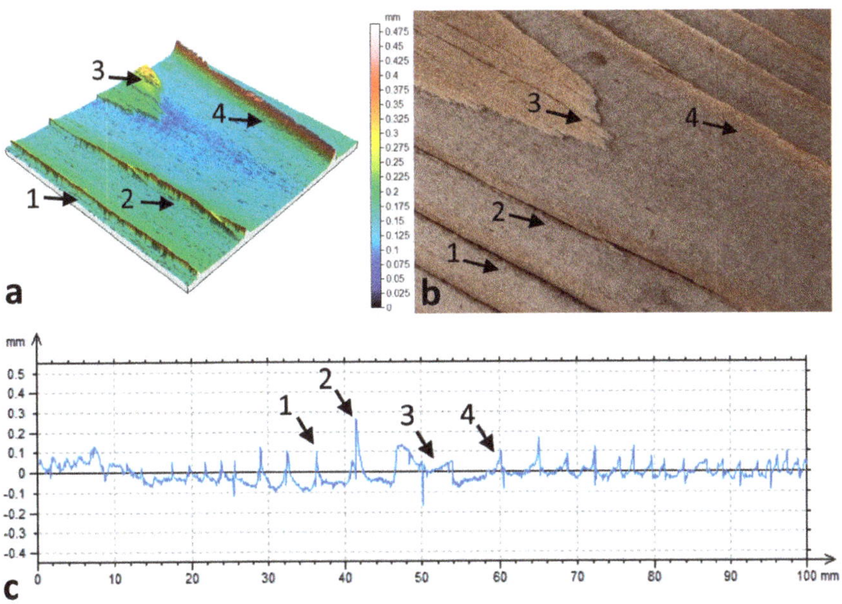

Figure 2. Shelling of growth rings in a flat western larch deckboard oriented pith-side-up and exposed to the weather for 7 months in Vancouver, BC, Canada. Four shelled growth rings including a shelled latewood tip are labelled, 1, 2, 3, and 4: (**a**) confocal profilometry height map showing four shelled growth rings; (**b**) photograph of four shelled growth rings matching the height map in Figure 2a; and, (**c**) confocal profilometry line scan showing the heights of the shelled growth rings in Figure 2a,b.

The three-dimensional (3D) profilometry height map and line scan of a weathered western larch deckboard with growth rings oriented pith-side-up shows the extent to which latewood projected above the wood surface, Figure 2a,c. For example, the numbered (1, 2, and 4) red coloured regions of the profilometry height map in Figure 2a correspond to the similarly numbered latewood bands in growth rings in Figure 2b. The heights of latewood bands in Figure 2a and the line scan Figure 2c indicate that one of the latewood bands (number 2) projected approximately 0.3 mm above adjacent earlywood. The latewood tip in Figure 2a,b (number 3) projected 0.15 mm above earlywood. The central area of Figure 2a,b is where a growth ring is parallel (0°) to the underlying surface. The heights of shelled latewood tend to be greatest in growth rings adjacent to this central region (Figure 2c), and then diminish in height further away from the central part of the deckboard. Latewood separated from the earlywood in the adjacent growth ring (inter-ring failure) rather than from earlywood in the same ring (intra-ring failure).

Table 2 shows profilometry measurements of the heights that shelled growth rings projected above the surface of each of the six western larch deckboard samples oriented pith-side-up and exposed to the weather. This table also includes the density of the parent boards and the numbers of growth rings per cm. The extent to which shelled growth rings projected above the surface of deckboard samples varied between the six boards, but there appeared to be no relationship between wood density and growth ring width and severity of shelling, as in Table 2.

Table 2. Wood properties and shelling of six different flat western larch deckboards oriented pith-side-up and exposed to the weather in Vancouver, BC, Canada for 7 months.

Board No.	Height Shelling (mm) [1]	No. Shelled Growth Rings [2]	Wood Density (g/cm^3)	Growth Rings/cm
1	0.10 (0.16–0.063)	6 (36)	0.517	9.0
2	0.12 (0.31–0.056)	19 (37)	0.431	5.0
3	0.18 (0.48–0.081)	14 (30)	0.456	6.0
4	0.08 (0.21–0.019)	32 (55)	0.467	6.0
5	0.05 (0.07–0.013)	12 (54)	0.442	7.5
6	0.19 (0.52–0.068)	17 (47)	0.488	8.5

[1] Average height of shelled growth rings. Max and min in parentheses; [2] Total number of growth rings in parentheses.

There was an inverse relationship between the height of shelled growth rings and the angle growth rings made to the surface of flat boards oriented pith-side-up, as in Figure 3. In other words, shelled growth rings with a smaller angle to the surface tended to project further from the surface than growth rings that were inclined at a higher angle to the surface. The regression between shelled growth ring height and growth ring angle was very significant ($p < 0.001$), but the correlation coefficient was low ($R^2 = 0.327$). This may be caused by our small sample size.

Shelling did not occur at the surface of flat weathered deckboards oriented bark-side-up, although checking was more pronounced at the surface of boards oriented bark-side-up compared to those oriented pith-side-up, Figure 4, in accord with the results of previous studies [5,8]. Many of the checks in boards oriented bark-side-up occurred at the interface of latewood and earlywood (Figure 4), but did not cause growth rings to project from the surface of the wood, as was observed in boards oriented pith-side-up. Latewood was raised slightly above the surface of adjacent earlywood possibly due to springback or differential swelling of latewood and/or increased degradation and erosion of earlywood compared to latewood [6,26].

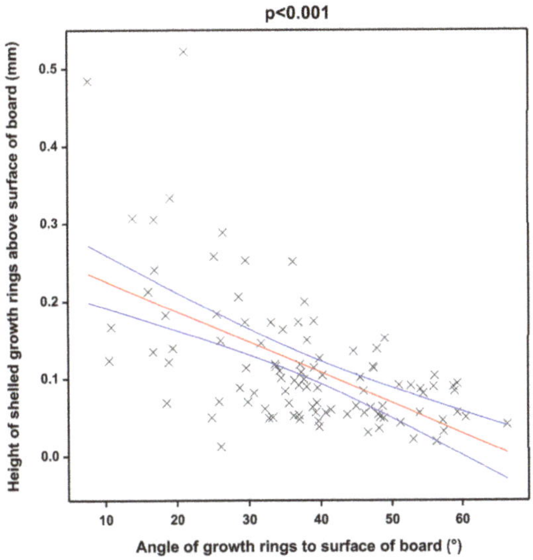

Figure 3. Linear regression (red line) of heights of latewood in shelled growth rings versus growth ring angle in flat western larch deckboards oriented pith-side-up and exposed to the weather for seven months in Vancouver, Canada; regression includes 95% confidence intervals (blue-lines).

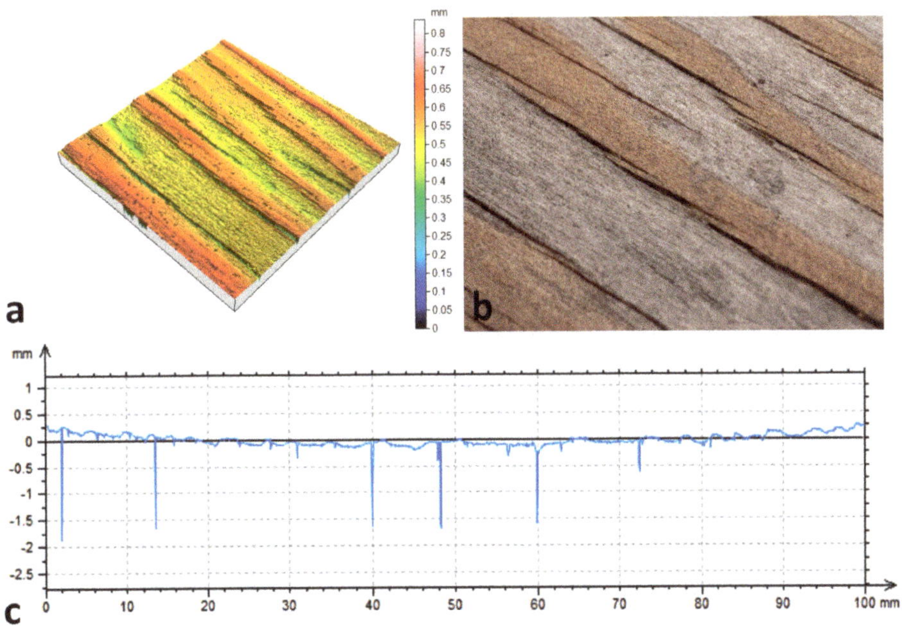

Figure 4. Appearance of a western larch deckboard with growth rings oriented bark-side-up and exposed to the weather for 7 months in Vancouver, Canada: (**a**) confocal profilometry height map showing surface checking and roughening of the wood surface; (**b**) photograph matching the image in Figure 4a; and, (**c**) confocal profilometry line scan showing variation in height of images in Figure 4a,b.

3.2. Shelling of Growth Rings in Flat ACQ-Treated Douglas Fir, Western Hemlock and White Spruce Deckboards

Confocal profilometry imaging of flat ACQ-treated western hemlock, white spruce, and Douglas fir deckboards oriented pith-side-up and exposed to the weather for 18 months showed that shelling in these species was due to inter-ring delamination of growth rings and projection of latewood above adjacent earlywood, as in Figure 5. In western hemlock and white spruce, latewood in shelled growth rings projected over 1 mm above the surface, and splitting of delaminated latewood occurred in western hemlock creating sharp splinters. The delamination of growth rings and projection of latewood above weathered deckboard surfaces was more pronounced in western hemlock and white spruce than in Douglas fir boards.

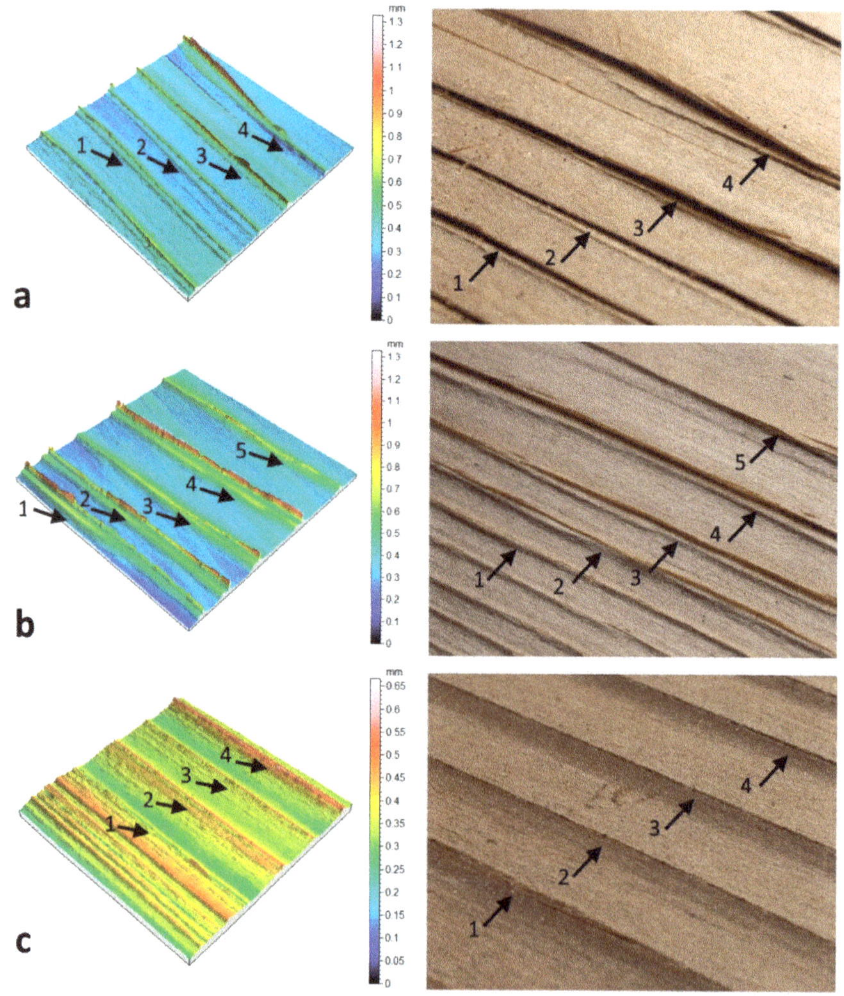

Figure 5. Confocal profilometry height maps showing shelling of growth rings in alkaline copper quaternary (ACQ) treated (**a**) western hemlock, (**b**) white spruce, and (**c**) Douglas fir deckboards with growth rings oriented pith-side-up and exposed to the weather for 18 months in Vancouver, BC, Canada.

Scanning electron microscopy was used to examine the interface between latewood and earlywood in shelled growth rings in Douglas fir, western hemlock, and white spruce deckboards, as in Figure 6. Images of shelled growth rings in all three species show a check between thicker-walled latewood cells and thinner-walled earlywood cells. In western hemlock, there was evidence that delamination of growth rings resulted from fracture of thin walled earlywood tracheids, as shown in Figure 6d,f. All three species showed evidence of photodegradation, for example, pit micro-checking (arrowed in Figure 6a,c) and colonization of the wood surface by mould fungi (arrowed in Figure 6d).

Figure 6. Scanning electron microscopy images of shelled growth rings in small samples from flat ACQ-treated Douglas fir, white spruce and western hemlock deckboards with growth rings oriented pith-side-up and exposed to the weather for 18 months in Vancouver, Canada: (**a**) Douglas fir sample showing separation of latewood (right) from earlywood. Note pit micro-checking in earlywood (arrowed bottom left); (**b**) white spruce sample showing separation of latewood (right) from earlywood; (**c**) western hemlock sample showing separation of latewood (right) from earlywood. Note pit micro-checking in earlywood (arrowed centre left); (**d**) western hemlock sample showing failure of earlywood cell wall (centre) where latewood separates from earlywood. Note mould in earlywood (arrowed bottom left); (**e**) a cross section through a shelled growth ring in western hemlock showing separation of latewood (left) from earlywood; and, (**f**) a cross section through a shelled growth ring in western hemlock showing failure of earlywood cell walls where latewood (left) separates from earlywood. Scale bars (**a–c**,**e**) = 100 μm; (**d**,**f**) = 20 μm.

3.3. Shelling of Growth Rings in Profiled Western Larch Deckboards

Weathering caused the separation of latewood and earlywood in some profile peaks in profiled deckboards with growth rings oriented pith-side-up. The separation of growth rings created lanceolate-shaped slivers (tips) of latewood that projected above the surface of deckboards, as in Figure 7. Figure 7a,c shows the extent to which the slivers projected above the rib of a profiled board. Shelling at the edges of growth rings was less obvious in profiled deckboards than in flat boards, even though latewood clearly separated from earlywood Figure 7b.

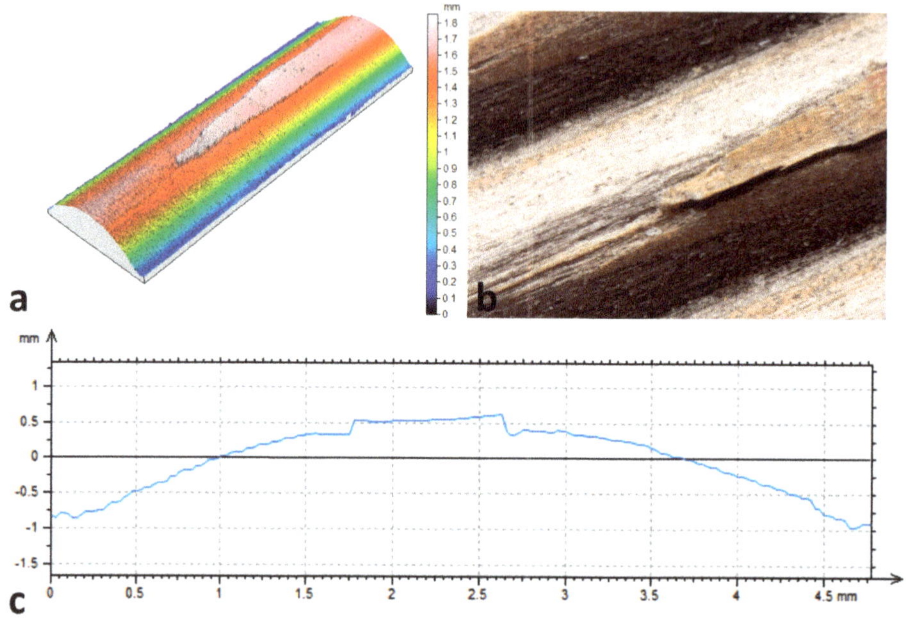

Figure 7. Shelling of a growth ring tip on the peak of a profiled (rib) western larch deckboard with growth rings oriented pith-side-up and exposed to the weather for seven months in Vancouver, Canada: (**a**) confocal profilometry height map of a shelled latewood tip; (**b**) photograph of a shelled latewood tip matching the image in Figure 7a; and, (**c**) confocal profilometry line trace showing variation in height of the latewood tip in Figure 7b.

Growth rings rarely separated (shelled) at the interface of latewood and earlywood at the peaks of profiled deckboards with growth rings oriented bark-side-up. However, latewood tips were slightly raised above the surface of adjacent earlywood, as can be seen in the profilometry height map and photograph in Figure 8. The extent to which latewood was raised above the surface of the adjacent earlywood was much smaller than that of shelled latewood tips in profiled boards oriented pith-side-up (compare Figures 7c and 8c).

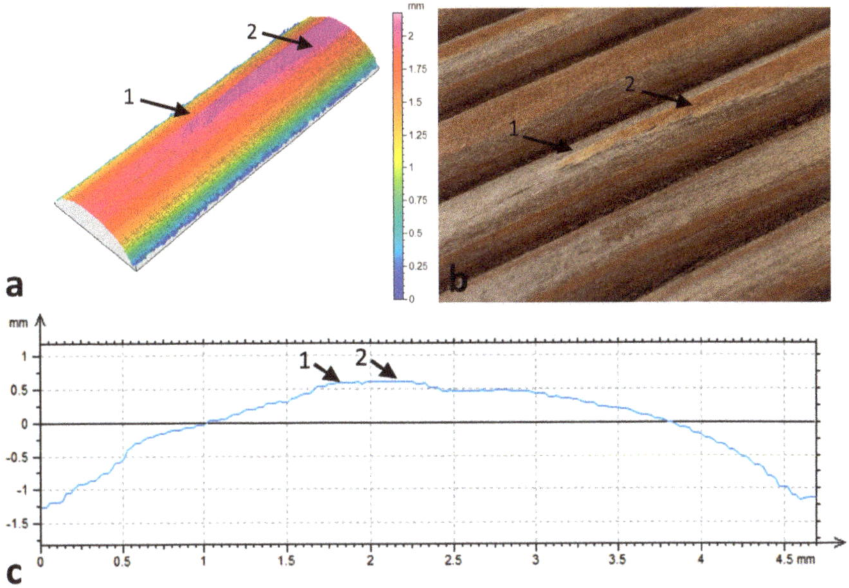

Figure 8. Tip of a growth ring raised above the peak of a profiled (short rib) western larch deckboard with growth rings oriented bark-side-up and exposed to the weather for seven months in Vancouver, Canada: (**a**) confocal profilometry height map of a latewood tip; (**b**) photograph of a latewood tip matching the image in Figure 8a; and, (**c**) confocal profilometry line trace showing variation in height of the arrowed rib in Figure 8b.

We counted the number of latewood tips that visibly projected above the surface of profile peaks in boards with different growth ring orientations (pith-side-up and bark-side-up) and profile geometries (rib, short rib, tall rib, ribble, and ripple). Analysis of variance indicated that the number of visible latewood tips at the surface of profiled and weathered deckboards oriented pith-side-up was significantly ($p < 0.001$) greater than that of similarly weathered boards oriented bark-side-up, as in Figure 9.

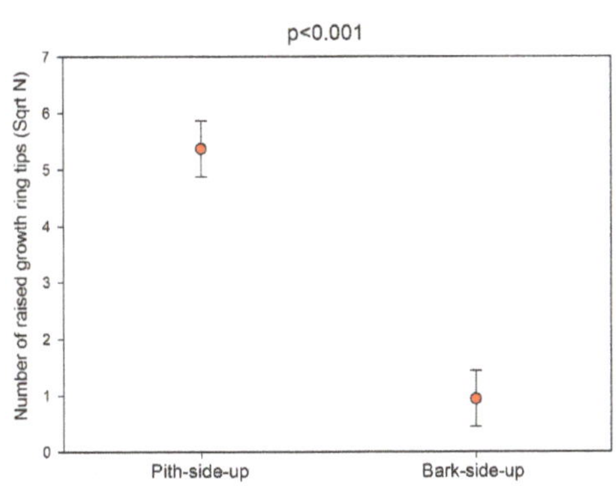

Figure 9. Effects of growth ring orientation on the numbers of latewood tips that projected above the surface of profiled western larch deckboards exposed to natural weathering for seven months in Vancouver, Canada. The results are averaged across boards with different profiles because there was no significant ($p > 0.05$) interaction of growth ring orientation and profile type on numbers of raised latewood tips. Error bars = Least significant difference, $p < 0.05$.

There was a small, but statistically significant ($p = 0.036$) effect of profile type (rib, ribble, and ripple, etc.) on the number of latewood tips that visibly projected above the surface of peaks in profiled boards, Figure 10, but the interaction of growth ring orientation *x* profile type was not significant ($p > 0.05$). Boards with a wavy ribble profile had significantly ($p < 0.05$) fewer visible latewood tips when compared to other profiled boards, except for boards with a wavy ripple profile (Figure 10).

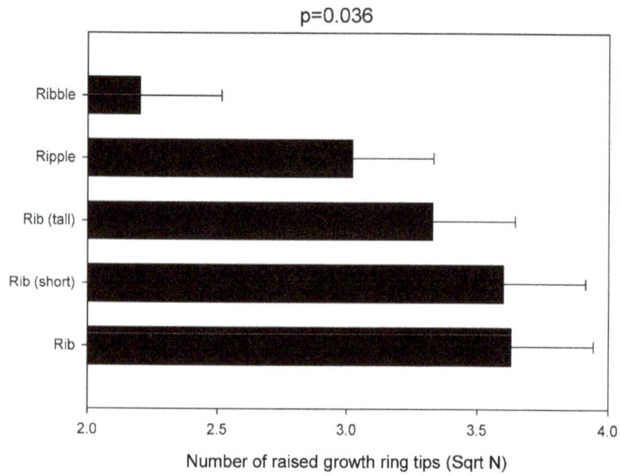

Figure 10. Effects of profile type on the numbers of latewood tips that projected above the surface of profiled western larch deckboards exposed to natural weathering for seven months in Vancouver, Canada. Results are averaged across boards with different growth ring orientations because there was no significant ($p > 0.05$) interaction of growth ring orientation and profile type on numbers of raised latewood tips. Error bars = Least significant difference, $p < 0.05$.

4. Discussion

Our results support our hypothesis that confocal profilometry can quantify shelling at the surface of weathered deckboards, but the methodology is not completely straight-forward. The surfaces of weathered deckboards contain features that span various scales. For example, in the *Z*-direction (height), wood cell walls and raised grain can project 10 to 50 µm from wood surfaces, respectively [16,17,27]. Weathering can lead to millimetres of erosion of wood [6], and these variable perturbations in the vertical heights of deckboard surfaces can be superimposed upon more structured surface features, such as machined profiles [21]. These multiscale features of wood surfaces have all been quantified using confocal profilometry, but probes with higher sensitivity and smaller sampling areas are required for features in the micron range as compared to those in millimeter range [16–19,27,28]. Hence, selective sampling of areas of deckboard surfaces showing shelling was required, particularly for profiled boards, to extract information on the maximum height of shelled growth rings, which was the parameter of greatest practical interest. When we took this approach, confocal profilometry was able to reveal a negative correlation between growth ring angle and severity of shelling in flat-sawn western larch boards oriented pith-side-up, and confirm previous suggestions in the literature that flat deckboards oriented pith-side-up are more susceptible to shelling than boards oriented bark-side-up [7,9–11]. Our results also accord with results of Davis, cited in Koehler that wood density and width of annual ring appear to have no effect on shelling, and also his suggestion that wood species differ in their susceptibility to shelling [9,10]. Further research is needed to fully explore species differences in susceptibility to shelling, and the influence of other factors on shelling, for example, contrast in density between earlywood and latewood [10], and the application of water-repellent finishes [29].

We showed, for the first time, that shelling occurs at the surface of profiled deckboards exposed to the weather, and we described the morphology of shelling that developed in profiled boards. The shelling of profiled boards was due to the separation of latewood tips on the peaks of profiles, and our results suggested that boards with a wavy profile were less susceptible to shelling than boards with rib profiles. One possible explanation for this observation is that boards with wavy profiles have narrower peaks (1.2 to 1.3 mm, Table 1), which could reduce the possibility of latewood tips occurring on the peaks of profiles, and subsequently separating from earlywood, when compared to rib profiles that have wider peaks (2.2 to 2.4 mm, Table 1). Shelling of profiled boards was more pronounced in deckboards oriented pith-side-up compared to those oriented bark-side-up, in accord with our observations of the shelling of flat (unprofiled) boards.

Our observations confirm that shelling at the surface of boards oriented pith-side-up occurs due to fracture at the boundary between the latewood of one growth ring and the earlywood of the subsequent growth ring [9,10]. It is at this point that there is the greatest contrast in density between latewood and earlywood tracheids [30] and, hence, the potential for maximum stress concentration due to differential swelling and shrinkage of latewood and earlywood. These stresses are responsible for delamination at growth ring boundaries according to Koehler [9]. Intra-wall rather than interfacial fracture was observed in earlywood cell walls at the growth ring boundary of a shelled growth ring in western hemlock. We suggest that fracture at the latewood-earlywood boundary allows for unrestrained swelling of latewood, causing it to curl away from the wood surface in the same direction as the curvature of growth rings. We did not observe similar shelling in boards oriented bark-side-up although latewood projected slightly above wood surfaces, possibly due to springback or differential swelling of latewood and/or increased erosion of adjacent earlywood. Checking occurred at the interface between latewood and earlywood in boards oriented bark-side-up, but extended radially into wood, rather than propagating via the boundary between latewood and earlywood. The mechanisms that are responsible for the pronounced difference in shelling of boards oriented pith-side-up versus those oriented bark-side-up are well explained by Koehler [9] and in an anonymous publication by the U.S. Forest Products Laboratory that draws upon Koehler's work [10]. Our observations accord with Koehler's explanations [9] for the effects of growth ring orientation on the shelling of flat-sawn boards.

Shelling is a serious defect in wood species with a pronounced contrast between earlywood and latewood density [9,10]. Shelling of western larch wood was mentioned by Johnson and Bradner [12], who stated that it is objectionable because of danger from splinters, and its potential to create a tripping hazard. Our results indicate that the simplest way of avoiding the occurrence of shelling at the surface of flat western larch deckboards, is to orient boards bark-side-up rather than pith-side-up. Shelling can also be minimized, as others have pointed out, by machining boards with sharp cutter knives [9,10], and using a water-repellent treatment to reduce surface changes in wood moisture content [29]. All of these recommendations are also relevant to profiled larch deckboards, which developed shelling when exposed to the weather. An additional measure that would help reduce shelling of profiled larch deckboards is to use a wavy rather than a rib profile. Profiled deckboards should be profiled on both sides to produce a balanced board that is less susceptible to cupping [8]. A profile on the pith-side of boards that is different to that on the bark-side would make it easier to orient deckboards correctly, although the sub-surface profile would need to be able to achieve its desired aim of reducing cupping. Further research is needed to develop such sub-surface profiles, which could include saw kerfs or grooving that are employed on the undersides of flooring and some commercial deckboards [21,31].

5. Conclusions

Non-contact confocal profilometry was able to quantify shelling of growth rings at the surface of flat and profiled softwood deckboards that were exposed to natural weathering. Using this technique, we have shown how the orientation of growth rings and the angle growth rings make to wood surfaces influences the severity of shelling in flat western larch deckboards exposed to natural weathering. We also show that profilometry can be used to quantify shelling in other wood species and suggest that it can provide additional insights into how the properties of wood or treatments, such as coatings and water repellents, influence shelling. Our results confirm that shelling occurs due to inter-ring delamination in deckboards with growth rings oriented pith-side-up. We show, for the first time, that shelling occurs at the surface of profiled deckboards oriented pith-side-up. We conclude that the shelling of flat and profiled boards can be reduced by orienting deckboards bark-side-up and, in the case of profiled boards, using a wavy profile rather than a hemispherical rib profile.

Author Contributions: L.H.L. and P.D.E. conceived and designed the experiments; L.H.L. performed all experimental work, except for scanning electron microscopy (P.D.E.). P.D.E. analyzed all data and wrote the first draft of the paper. Both authors discussed and commented on the results and contributed to the final submitted manuscript. All authors have read and agree to the published version of the manuscript.

Funding: This research was funded by Natural Science and Engineering Research Council of Canada (NSERC) Collaborative Research and Development Grant (CRDPJ 485007-15).

Acknowledgments: We thank Steve Brown, Jarett Pereboom, Terry Cunning, Arne Flaten, Jeff Hunt, Ken Johnstone Jr. and Dan Price from Tolko for a special order of western larch lumber; Brandon Chan, Kenneth J. Cheng, Pablo Chung, Lawrence Gunther, Sina Heshmati, Joseph Doh Wook Kim and Mohammad Sadegh Mazloomi from UBC for helping to machine and characterize deckboard samples; Hua Chen and Frank Brink of the Centre for Advanced Microscopy at The Australian National University (ANU) for assistance with scanning electron microscopy; Canadian Foundation for Innovation, BC Knowledge Development Fund, FPInnovations and Tolko industries for in-kind support. P.D.E. thanks Viance, Tolko, FPInnovations, Faculty of Forestry (UBC) and the Government of British Columbia for their support of his BC Leadership Chair at the University of British Columbia, and The Australian National University (ANU) for an Honorary Professorship in the Department of Applied Mathematics, Research School of Physics at the ANU.

Conflicts of Interest: The authors declare no conflict of interest, and none of the individuals or organizations acknowledged above were involved in the design of our experiment, collection, and interpretation of data or the writing of this paper.

References

1. Feist, W.C.; Hon, D.N.S. Chemistry of weathering and protection. In *The Chemistry of Solid Wood*; Advances in Chemistry Series; Rowell, R.M., Ed.; American Chemical Society: Washington, DC, USA, 1984; Volume 207, Chapter 11; pp. 401–405.
2. Lie, S.K.; Vestøl, G.I.; Høibø, O.; Gobakken, L.R. Visual appearance of unpainted wood: Mould coverage, lightness and uniformity. *Int. Wood Prod. J.* **2019**, *10*, 9–15. [CrossRef]
3. Christy, A.G.; Senden, T.J.; Evans, P.D. Automated measurement of checks at wood surfaces. *Measurement* **2005**, *37*, 109–118. [CrossRef]
4. Schniewind, A.P. Mechanism of check formation. *For. Prod. J.* **1963**, *13*, 475–480.
5. Yata, S. Occurrence of drying checks in softwood during outdoor exposure. In *High-Performance Utilization of Wood for Outdoor Uses*; Imamura, Y., Ed.; Wood Research Institute, Kyoto University: Kyoto, Japan, 2001; pp. 65–70.
6. Williams, R.S.; Knaebe, M.T.; Feist, W.C. Erosion rates of wood during natural weathering: Part II Earlywood and latewood erosion rates. *Wood Fiber Sci.* **2001**, *33*, 43–49.
7. Williams, S.; Knaebe, M. *The Bark Side/Pith Side Debate. The Finish Line*; U.S. Forest Service, Forest Products Laboratory: Madison, WI, USA, 1995; pp. 1–2.
8. Heshmati, S.; Leung, L.H.; Mazloomi, M.S.; Kim, J.D.W.; Evans, P.D. Technical note: Positive effects of double-sided profiling on the cupping and checking of ACQ-treated Douglas fir, western hemlock, and white spruce deckboards exposed to natural weathering. *Wood Fiber Sci.* **2020**, *52*, 243–253. [CrossRef]
9. Koehler, A. Raised grain—Its causes and prevention. *Southern Lumberman.* **1929**, *137*, 210M–210O.
10. Anonymous. *Raised, Loosened, Torn, Chipped, and Fuzzy Grain in Lumber*; FPL-099; U.S. Forest Service, Forest Products Laboratory: Madison, WI, USA, 1965; pp. 1–15.
11. Williams, S.; Knaebe, M. The finish line: Practical facts on wood finishing from FPL. *Am. Paint Coat. J. Conv. Dly.* **1997**, 26.
12. Johnson, R.P.A.; Bradner, M.I. *Properties of Western Larch and Their Relation to Uses of the Wood*; Technical Bulletin 282; United States Department of Agriculture: Washington, DC, USA, 1932; pp. 1–93.
13. Kavanaugh, C. Vinyl Siding Tops Cladding List for 25 Years. Plastics News. July 2019. Available online: https://www.plasticsnews.com/news/vinyl-siding-tops-cladding-list-25-years (accessed on 17 June 2020).

14. Kavanaugh, C. Composite Decking Making Market Gains. Plastics News. July 2019. Available online: https://www.plasticsnews.com/news/composite-decking-making-market-gains (accessed on 17 June 2020).
15. Nakamura, G.-I.; Takachio, H. An experiment on the roughness and stability of sanded surface. *Mokuzai Gakkaishi* **1961**, *7*, 41–45.
16. Landry, V.; Blanchet, P.; Cormier, L.M. Water-Based and solvent-based stains: Impact on the grain raising in yellow birch. *Bioresources* **2013**, *8*, 1997–2009. [CrossRef]
17. Evans, P.D.; Cullis, I.; Kim, J.D.W.; Leung, L.H.; Hazneza, S.; Heady, R.D. Microstructure and mechanism of grain raising in wood. *Coatings* **2017**, *7*, 135. [CrossRef]
18. Singh, A.; Dawson, B.; Turner, J.; Rickard, C. *Anatomical Explanation for Grain Raising in Machined Radiata Boards*; Wood Processing Newsletter 41; Scion: Rotorua, New Zealand, 2008; pp. 3–6.
19. Dawson, B.S.W.; Singh, A.P.; Ward, J.; Smolic, T.; Singh, A.; Hands, K.D. Texture of wooden surfaces before and after coating. *Surf. Coat. Aust.* **2006**, *43*, 14–18.
20. Böttcher, P. Einfluß verschiedenartiger oberflächenprofilierungen an holz auf die veränderung der wetterbeständigkeit. *Holz Roh Werkst.* **1977**, *35*, 247–251. [CrossRef]
21. Cheng, K.J.; Evans, P.D. A note on the surface topography of profiled wood decking. *Aust. For. J.* **2016**, *79*, 147–152. [CrossRef]
22. Jamali, A.; Evans, P.D. Plasma treatment reduced the discoloration of an acrylic coating on hot-oil modified wood exposed to natural weathering. *Coatings* **2020**, *10*, 248. [CrossRef]
23. Heady, R.D.; Banks, J.G.; Evans, P.D. Wood anatomy of Wollemi pine (*Wollemia nobilis*, Araucariaceae). *IAWA J.* **2002**, *23*, 339–357. [CrossRef]
24. Cheng, K.J.; Evans, P.D. Manufacture of profiled amabilis fir deckboards with reduced susceptibility to surface checking. *J. Manuf. Mater. Process.* **2018**, *2*, 7. [CrossRef]
25. Williams, L.J.; Hervé, A. Fisher's least significant difference (LSD) test. In *Encyclopedia of Research Design*; Salkind, N., Ed.; SAGE: Thousand Oaks, CA, USA, 2010; p. 6.
26. Kataoka, Y.; Kiguchi, M.; Fujiwara, T.; Evans, P.D. The effects of within-species and between-species variation in wood density on the photodegradation depth profiles of sugi (*Cryptomeria japonica*) and hinoki (*Chamaecyparis obtusa*). *J. Wood Sci.* **2005**, *51*, 531–536. [CrossRef]
27. Jamali, A.; Evans, P.D. Etching of wood surfaces by glow discharge plasma. *Wood Sci. Technol.* **2011**, *45*, 169–182. [CrossRef]
28. Hazneza, S.; Evans, P.D. End-Grain erosion of Douglas fir wood during natural weathering. *Int. Wood Prod. J.* **2016**, *7*, 3–11. [CrossRef]
29. Schuttner, S. *Building and Designing Decks*; The Taunton Press Inc.: Newtown, CT, USA, 1998; p. 164.
30. Björklund, J.; Seftigen, K.; Schweingruber, F.; Fonti, P.; von Arx, G.; Bryukhanova, M.V.; Cuny, H.E.; Carrer, M.; Castagneri, D.; Frank, D.C. Cell size and wall dimensions drive distinct variability of earlywood and latewood density in Northern Hemisphere conifers. *New Phytol.* **2017**, *216*, 728–740. [CrossRef] [PubMed]
31. Nystrom, R. Board for Use in Constructing a Flooring Surface. U.S. Patent 5,474,831, 12 December 1995.

coatings

Article

The Impact of Fungicides, Plasma, UV-Additives and Weathering on the Adhesion Strength of Acrylic and Alkyd Coatings to the Norway Spruce Wood

Ladislav Reinprecht [1,*], Radovan Tiňo [2] and Marek Šomšák [3]

1 Department of Wood Technology, Faculty of Wood Sciences and Technology, Technical University in Zvolen, T. G. Masaryka 24, SK-960 01 Zvolen, Slovakia
2 Faculty of Chemical and Food Technology, Slovak Technical University, Radlinského 9, SK-812 37 Bratislava, Slovakia; radko.tino@stuba.sk
3 Agrotechnical Secondary Vocational School, SK-045 01 Moldava nad Bodvou, Slovakia; somsak.marek@gmail.com
* Correspondence: reinprecht@tuzvo.sk

Received: 28 October 2020; Accepted: 16 November 2020; Published: 19 November 2020

Abstract: The adhesion strength between the transparent acrylic or alkyd coatings and the Norway spruce (*Picea abies* Karst L.) wood was determined by EN ISO 4624 and analyzed concerning four variables: (a) fungicidal pre-treatment of wood with boric acid or benzalkonium chloride, (b) cold plasma modification of wood surfaces, (c) presence of hindered amine light stabilizer (HALS) or hydroxyphenyl-benzotriazoles (BTZ) in the role of UV-additives in coatings, and (d) weathering of coated wood—lasting 1 week in Xenotest by a modified EN 927-6, or 14, 28 and 42 weeks outdoors at 45° by EN 927-3. In the un-weathered state, the adhesion strength was positively affected by the initial plasma modification of wood surfaces, more evident with the application of acrylic water-borne coatings. On the contrary, the adhesion strength was not influenced by the fungicidal pre-treatment of wood and by the UV-additive's presence in coatings. The adhesion was negatively affected by weathering—exponentially outdoor—irrespective of the fungicidal pre-treatment of wood, the plasma modification of wood surfaces, the coating type, and the presence of UV-additive in coatings.

Keywords: spruce wood; fungicides; plasma; coatings; UV-additives; weathering; adhesion

1. Introduction

In outdoor expositions, unprotected wood surfaces are easily attacked by sunlight. The penetration depth of UV-radiation into wood is approximately 75 micrometers and of the visible light 125–500 micrometers [1]. However, free radicals created from the initially photodegraded lignin and hemicellulose macromolecules can further damage wood components up to a depth of approximately 2000 micrometers. Surfaces of wet wood are synchronously susceptible to deterioration processes by bacteria and molds causing color defects and also health problems for people [2].

Coatings recommended for wooden products that are exposed exteriorly, e.g., windows, pergolas, façades, or terrace boards, must protect them against sunlight, water, and microorganisms. This means that the commercial acrylic, alkyd, polyurethane, epoxy, or other coating types should contain suitable additives—pigments, UV-additives, hydrophobic substances, and biocides.

The pigments and UV-additives absorb sunlight, decrease destructive effects of sunlight, or inhibit its transfer through coatings into wood [3–6]. The hydrophobic substances, e.g., natural and synthetic waxes, increase a contact angle between the coated wood and the water drops and decrease the penetration of water through the coating into wood [7]. The environmentally acceptable bactericides and fungicides protect coatings and wood against bio-deteriorations [8].

The transparent, unpigmented coatings used in exteriors should contain effective UV-additives that preserve the macromolecules of coatings from radical depolymerization reactions leading to their destruction and protect the painted wood from color changes and other aesthetical defects [9]. Today, the following substances are applied as UV-additives in coatings: (a) UV absorbers such as 2-hydroxyphenyl-s-triazines [5,10], 2-(2-hydroxyphenyl)-benzotriazoles (BTZ) [3,5], or 2-hydroxyphenyl-benzophenones [11], (b) UV blockers-screeners such as zinc oxide (ZnO), titanium dioxide (TiO_2) or cesium dioxide (CsO_2), usually used in the form of nanoparticles [3,5,6,12–14], (c) hindered amine light stabilizers (HALS) [15,16], (d) sometimes also with other photo-stabilization mechanisms, for example, imidized nanoparticles which also have a hydrophobic effect [17], or lignin stabilizers such as succinic anhydride in combination with epoxidized soybean oil [11]. Unfortunately, the efficiency of UV-additives in coatings does not always last long-term [18,19].

Coatings for wood products exposed outdoors are increasingly modified with environmentally acceptable organic bactericides and fungicides, for example 3-iodo-2-propinyl-N-butyl-carbamate (IPBC), propiconazole, tebuconazole, or quaternary ammonium compounds (QACs), or less frequently with inorganic biocides, for example nano ZnO, or nano silver [7,8,20,21]. These biocides, as well as other ones, can also be used for wood pre-treatment before the application of coatings [22].

The functional, protective, and decorative properties of coatings used for the treatment of wood surfaces exposed exteriorly should be sufficient and long-term [23]. However, the adhesion strength between coatings and wood surfaces usually worsens over time in relation to: (1) the type and intensity of environmental factors, i.e., the sun-irradiation, water precipitation, wind, emission of carbon blacks and aggressive chemicals, (2) the presence of bacteria, molds or other pests, (3) the used wood species with specific surface characteristics and its physical, chemical or biological pre-treatments before painting with coating, and (4) the used coating type, and the technology of its application.

Individual wood species have specific surface characteristics—wettability, free surface energy, porosity, roughness, pH-value, resistance to bacteria, molds, decaying fungi, etc. [24–29]. All surface characteristics of wood are directly connected with its geometrical, morphological, anatomical, and molecular structure [30]. Differences in the wood surface texture of various wood species or of specimens of the same wood species are given by (a) the geometrical level, such as the radial or tangential surface, knots, roughness influenced by machining, etc., (b) the morphological and anatomical levels, such as the diameter of cell elements "fibers, vessels, rays, ... " and their lumens, etc., and (c) the molecular level, such as the type and amount of hydroxyl and other polar functional groups, crystallinity of cellulose, polarity of lignin–polysaccharide components and various extractives, migration of extractives, etc. The surface texture of wood significantly influences the wettability and penetration processes of liquid coatings into wood as well as the adhesion of created coating films to wood.

However, the wood species, its roughness, and other surface characteristics are not always the most important factors that affect the adhesion strength, as the role of the coating, and similarly of the glue, is usually more important [31,32]. As already mentioned in the previous paragraphs, pigmented coatings are preferentially recommended for exterior usage, while the transparent coatings can only be used after adjusting them with effective and stable UV-additives, hydrophobic and antimicrobial substances. Kúdela and Liptáková [33] and several other researchers identified the "coating—wood" interface as a very important parameter for overall stability and the lifetime of coatings. The adhesion strength between coatings and wood surfaces can be improved by wood pre-treatment with chemical penetrating systems, as well as by wood modification with specific physical methods.

Modification of wood surfaces with plasma is a well-known physical method for their activation before the application of coatings or glues [34–37]. The plasma barrier discharge serves to activate wood surfaces prior to the application of coatings to ensure a higher adhesion strength and increase the service lifetime of coated or glued woods. Plasma can improve the surface properties of wood, wooden composites, and some other materials [37,38]. Plasma forms a thin protection layer on a wooden surface, which is effective against the sun, water, and biological influences. Several experiments have shown that plasma can change the chemical composition of wood surfaces, thereby also changing their wettability together with improving the adhesion strength of coatings and glues to wood [35,37–44]. The plasma discharge in wood surfaces activates polysaccharides, lignin and extractives and in the presence of air generates new hydroxyl, carbonyl, carboxyl, peroxide or ether functional groups and radicals, and simultaneously liberates reactive intermediates, such as O_2^+, $^1O_2$, O_3, O^+, ionized ozone, free electrons, N, CO_2, excited states of N_2, etc. [37,45]. These substances react with the wood surface and provide a convenient resource for its activation and purposeful alteration of its wettability [46].

In presence of ozone, wood components are transformed into smaller and also more polar molecules. Thus, newly formed compounds in wood surfaces can form new chemical bonds with the surrounding lignin–polysaccharide components of wood and also with molecules of the used coating system. Following this, an assumption can be pronounced—newly bonded joints in the plasma modified wood will ensure better adhesion of the coating to its surface with a positive effect on the final quality and lifetime of the wooden product. In addition, the surface modification of wood by plasma carried out in the air, with the production of ozone, is an effective sterilizing method for the killing of bacteria or molds [34,42].

In the last few years, the cold low-temperature plasma (LTP) is increasingly used for the surface modification of wood, wood composites, wood–plastic composites, and other technical materials, because the degrading thermal effect of the cold plasma on these materials is neglected [38,47]. LTP does not require high temperatures that are needed in the preparation of ThermoWoods (thermally modified wood at 180–220 °C), and it is effective without the presence of activating chemical agents that otherwise maximize the benefits of subsequently applied natural or synthetic coatings [48]. In addition, the reaction time of LTP with wood components is relatively short, usually only from a few seconds to several minutes, additionally, the process is performed in a dry environment, and no by-products are produced. The surface of the plasma modified wood becomes more polar, more hydrophilic, and with better wettability with water-based coating systems, for example with acrylic water dispersions. However, gradually, over a period of time, usually after 7 to 30 days of exposure to plasma, the wood surfaces become chemically inert and even more hydrophobic [45]. The hydrophobic character of pinewood surface was also achieved by Moghaddam et al. [49], who created a thin transparent superhydrophobic layer, consisting of nanoparticulate TiO_2 and the following deposition of plasma polymerized organic substances, effective for controlling its wettability and water repellency.

Generally, the permanent quality of painted wood is connected with its color stability, resistance to the creation of cracks, to thickness reduction and to biological attacks, as well as the good and long-term adhesion strength of the coating system with the wood surface. In exterior exposures, the durability of wood surfaces painted with transparent coatings are often limited to just their adhesion strength to wood substrate. Adhesion of coatings to wood is commonly susceptible to weathering conditions, but less information exists about the effect of wood pre-treatment with fungicides and plasma and also about the presence of UV-additives in coatings. The Norway spruce is a common and very important wood species, mainly in Central Europe, used for industrial and building structures, bridges, furniture, musical instruments, sport equipment and other uses. From this point of view, research aimed at improving its surface properties was performed in this work.

The basic aim of this work was to examine changes in the adhesion strength between transparent acrylic and alkyd coatings and the Norway spruce wood due to weathering, in parallel evaluating some other impacts, including the presence of HALS or BTZ UV-additives in coatings, the fungicidal pre-treatment of wood with boric acid or benzalkonium chloride and the plasma modification of wood surfaces.

2. Materials and Methods

2.1. Wood

Top surfaces of the naturally dried and planed Norway spruce (*Picea abies* Karst L.) boards were gradually ground along the grain on a belt sander with 80 grit and 120 grit sandpapers, and then freed from wooden dust with compressed air. Following this, the samples: (a) with a dimension of 80 mm × 45 mm × 8 mm (axial × radial × tangential) for the artificial weathering test, and (b) with a dimension of 375 mm × 78 mm × 20 mm (axial × radial × tangential) for the outdoor weathering test were prepared from boards. Selected samples, i.e., without biological damages, juvenile wood, and growth inhomogeneities, were conditioned at a temperature of 20 ± 2 °C and a relative air humidity of 65 ± 5% achieving their equilibrium moisture content of 12 ± 2%. The density of the spruce samples used in the experiment ranged from 450 to 475 kg·m^{-3}. The roughness parameters *Ra* (arithmetic mean deviation) and *Rz* (arithmetic mean of the heights and depressions of the profile at the basic length), determined for twenty selected samples perpendicular to the grain using the profilometer Surfcom 130A (Carl Zeiss, Jena, Germany) in accordance with the standard EN ISO 4287 [50], ranged in the intervals of 6.04–7.42 μm and 50.2–57.6 μm.

2.2. Fungicidal Pre-Treatment of Wood

For the fungicidal pre-treatment of spruce samples we used 2 wt.% water solutions of boric acid (H_3BO_3) (Funchem, Czech Republic) and benzalkonium chloride (BAC) belonging to QACs, synonym: alkylbenzyldimethylammonium chloride (Sigma Aldrich, Germany). Parameters of the soaking technology were as follows: a pressure of 100 kPa, a temperature of 30 °C, and an immersion time of 10 min. Spruce samples pre-treated with fungicides were then again conditioned achieving their equilibrium moisture content of 12 ± 2%.

2.3. Plasma Modification of Wood Surfaces

The top surfaces of spruce samples were modified with atmospheric low-temperature/cold plasma in laboratories of the Slovak Technical University in Bratislava, Slovakia, using the "Atmospheric Discharge with Runway Electrons" (ADRE) device (EST, Tomsk, Russia). The plasma modification was performed for one half of each group of spruce samples—the native, H_3BO_3 pre-treated, and BAC pre-treated. The following operating conditions were set in the ADRE device: a power of 1.4 kW, a frequency of 2000 Hz, and an exposure time of 60 s (Figure 1).

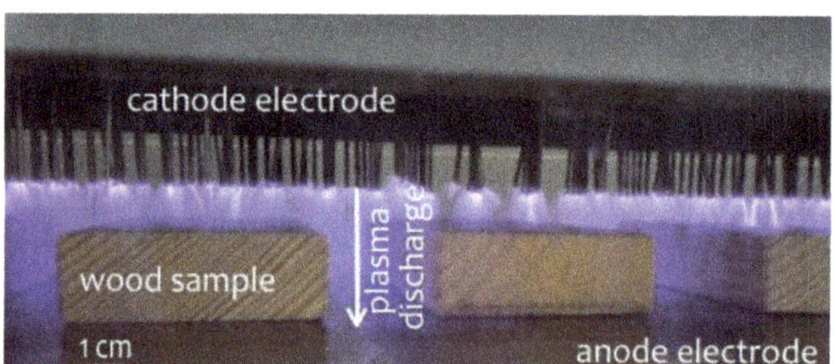

Figure 1. Plasma modification of the top surfaces of spruce samples in the "Atmospheric Discharge with Runway Electrons" (ADRE) device.

The initial contact angles γ_0 of redistilled water with the top surfaces of the natural (fungicidally un-pre-treated) spruce samples were measured using the SEE System (Advex Instruments, Brno, Czech Republic) instrument, always for 10 replicates. The contact angle γ_0 for the native unmodified spruce samples was 55° ± 2.3°, and after the plasma modification it actually decreased to

13° ± 1.6° as a result of the polar functional groups created in wood surfaces; however, after 24 h, it partly increased to 20° ± 5.5° and after 72 h even to 35° ± 6.5°—these tendencies are in accordance with [45].

2.4. Coatings and UV-additives

In experiment two types of transparent coatings synthesized by the company Chemolak Smolenice Ltd., Slovakia were used: (a) the acrylic water dispersion with a solids content of 28.6% (density 1.05 g/cm^3), and (b) the alkyd in non-aromatic petrol with a solids content of 54.9% (density 0.91 g/cm^3). Two UV-additive types were added into these coatings in the amounts of 0, 0.25, 0.5, and 1 wt.%: (a) the hindered amine light stabilizer (HALS) containing bis-(1,2,2,6,6-tetramethyl-4-piperidinyl)sebacate and methyl-(1,2,2,6,6,-tetramethyl-4-piperidinyl) sebacate: commercial product EVERSORB 93, (Everlight Chemical Industrial Corporation, Taipei, Taiwan), and (b) the UV absorber containing three types of hydroxyphenyl-benzotriazole (BTZ): commercial product EVERSORB 80 (Everlight Chemical Industrial Corporation, Taipei, Taiwan).

2.5. Painting of Wood Surfaces with Coatings

Before painting, the spruce samples were conditioned for 24 h at 20 ± 2 °C after plasma modification. This downtime was chosen exactly because with prolongation significant chemical changes in the plasma modified wood surfaces can occur, following by decreases in their hydrophilicity and wettability (see Section 2.3), [45,51], and also certainly in their adhesion with coatings. The pneumatic spraying technology was used for the coating's application in three layers—each layer in an amount of 150 ± 10 g per m^2 of the wood surface. The second layer and the third layer of coatings were implemented after a 24 h intermission needed for the drying and curing of the previous layer applied to the wood surface and for its subsequent sanding with 240 grit sandpaper.

2.6. Weathering of Coated Wood

The 1 week artificial weathering of coated spruce samples was performed in the Q-SUN Xe-1-S Xenotest (Q-Lab Corporation, Westlake, OH, USA). It took place in accordance with the modified version of standard EN 927-6 [52], i.e., 24 h conditioning of samples at 45 °C and then 48 subcycle steps—each lasting 3 h (2.5 h UV-radiation and then 0.5 h water spraying). These changes to [52] existed: Xenon lamps instead of fluorescent UV lamps; irradiance at 340 nm set to 0.55 W·m^{-2}·nm^{-1} instead of 0.89 W·m^{-2}·nm^{-1}; the temperature on the black panel at 50 °C instead of 60 °C.

The outdoor weathering of coated samples took place from May 12 2018 for periods of 0, 14, 28, and 42 weeks. It was performed in accordance with the standard EN 927-3 [53], at a slope of 45° in metal stands oriented to the South, located in the industrial town of Zvolen, Slovakia, at 300 m above sea level. The testing area is characterized by many foggy days, smog, and high-temperature differences between summer (to 35 °C) and winter (to −25 °C). Mean climatic conditions in this area were as follows: a temperature of 9.4 °C; a relative air humidity of 83%; water precipitation of 700 mm/year; and sun irradiation of 1100 kWh/m^2.

2.7. Adhesion between Coatings and Wood

The adhesion strength between the individual coating types and the individually pre-treated/modified spruce samples was determined with the pull-off test for adhesion in accordance with the standard EN ISO 4624 [54], using the PosiTest AT-M Adhesion Tester instrument (DeFelsko, Ogdensburg, NY, USA). A bond connection between the surface of the coated spruce sample and the steel-roller dolly with a diameter of 20 mm was made by two-component epoxy resin. The pull-off test tensile strength method measures the tensile force perpendicular to the phase interface "coating—wood" system, requiring the tearing off of the steel-roller from the coated spruce sample, at which the failures could occur either in the weakest interface "coating—wood" (usually typical in this work), or in the weakest component "coating" or "wood".

2.8. Statistical Analysis

The measured data of the adhesion strength were for individual groups of coated spruce samples evaluated on the basis of mean values and standard deviations. The effects of the HALS and BTZ UV-additives, added to coatings in concentrations (C) from 0 to 1 wt.%, were analyzed by the linear correlation "Adhesion = a + b × C". The effect of the outdoor weathering prolongation (τ) from 0 to 42 weeks was analyzed by the exponential correlation "Adhesion = a + b × exp (k × τ)". Within all correlations were determined the coefficients of determination R^2.

3. Results and Discussion

3.1. Adhesion Evaluated at Artificial Weathering

From the first experiment, connected with the artificial weathering of coated spruce samples, it is evident that the adhesion strength of the acrylic and alkyd coatings to the Norway spruce wood surfaces was not significantly influenced by the presence of the HALS and BTZ UV-additives in coatings in a range of concentration (C) from 0 to 1 wt.%. For example, for samples modified with plasma, the linear correlation "Adhesion = a + b × C" had in all cases a very small coefficient of determination R^2 from 0.001 to 0.151 (Table 1). A similar result was obtained for the natural (plasma un-modified) samples.

Table 1. No effect of the hindered amine light stabilizer (HALS) and hydroxyphenyl-benzotriazoles (BTZ) used in the role of UV-additives on the adhesion strength between the coating (acrylic or alkyd) and the plasma modified Norway spruce wood—evaluation at none and 1 week artificial weathering—confirmed by the small coefficients of determination R^2 from 0.001 to 0.151 for the linear correlations "Adhesion = a + b × C".

UV-Additive C (wt.%)	Adhesion between Coatings and Plasma Modified Wood (MPa)			
	None Weathering		1-Week Weathering in Xenotest	
	Acrylic	Alkyd	Acrylic	Alkyd
HALS				
0	2.90 (0.39)	2.92 (0.29)	2.13 (0.31)	2.43 (0.28)
0.25	3.20 (0.44)	2.70 (0.25)	2.62 (0.29)	2.01 (0.24)
0.5	2.80 (0.31)	2.43 (0.38)	2.35 (0.37)	2.20 (0.34)
1.0	2.93 (0.38)	2.47 (0.42)	2.30 (0.41)	2.32 (0.24)
"Adhesion = a + b × C"				
a	2.99	2.84	2.29	2.27
b	−0.08	−0.47	0.09	−0.03
R^2	0.002	0.105	0.032	0.027
BTZ				
0	2.90 (0.39)	2.92 (0.29)	2.13 (0.31)	2.43 (0.28)
0.25	2.77 (0.55)	2.73 (0.31)	2.33 (0.30)	2.30 (0.28)
0.5	2.74 (0.48)	2.70 (0.32)	2.37 (0.47)	2.43 (0.42)
1.0	3.03 (0.63)	2.97 (0.24)	2.47 (0.41)	2.27 (0.22)
"Adhesion = a + b × C"				
a	2.79	2.77	2.23	2.38
b	0.16	0.13	0.29	−0.08
R^2	0.002	0.001	0.151	0.023

Notes: the mean value is from 6 measurements, i.e., from 3 replicates with a different mode of fungicidal pre-treatment (none, H_3BO_3, or benzalkonium chloride (BAC)), testing each replicate on 2 places. The standard deviation is in parentheses. The coefficient of determination R^2 of the linear correlation (Adhesion = a + b × C) was determined from 24 values, i.e., 4 concentrations (C) of the UV-additive in the coating by 6 measurements.

The decrease in the adhesion strength between coatings and spruce samples due to the 1 week artificial weathering in Xenotest is documented by a drop in the parameter "a" in the linear correlations from 2.99–2.77 MPa to 2.38–2.23 MPa (Table 1). Analyzing the un-weathered samples, the acrylic and alkyd coatings—containing various amounts of the HALS or BTZ UV-additives—had a comparable adhesion strength with the spruce samples, as is documented by comparable values of the parameter "a" in the linear correlations, i.e., 2.99 or 2.79 MPa for acrylic coatings and 2.84 or 2.77 MPa for alkyd coatings (Table 1).

The impact of the HALS and BTZ UV-additives on the adhesion strength between coatings and plasma modified spruce wood was also analyzed as by the complex linear correlations "Adhesion = a + b × C" (Figure 2). For both UV-additive types, a comparable parameter "a" (2.60 or 2.54 MPa) was determined, together with a very small and opposite parameter "b" (−0.123 or 0.122 MPa), and with very small values of the coefficient of determination "$R^2 = 0.01$" and $p > 0.1$. These results confirm that the adhesion strength of the acrylic and alkyd coatings to the Norway spruce wood samples was not influenced by the presence of the UV-additives in the coatings.

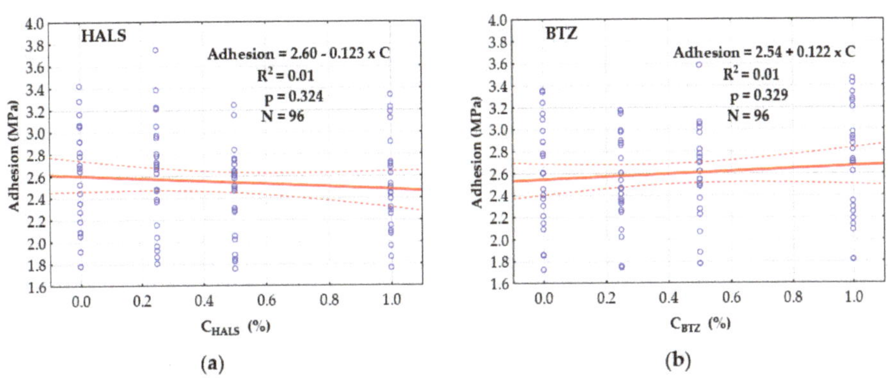

Figure 2. No effect of the UV-additives (**a**) HALS and (**b**) BTZ on the adhesion strength between the coatings and the plasma modified Norway spruce wood—confirmed by the small coefficient of determination R^2 equal to 0.01 and $p > 0.1$ for the linear correlations. Notes: the summary effect of UV-additives at the defined concentrations (C = 0, 0.25, 0.5, or 1 wt.%) is from 24 data measured for 6 replicates of 2 coating types "acrylic and alkyd" and 2 weathering modes "none and 1 week in Xenotest". The coefficient of determination R2 of the linear correlation (Adhesion = a + b × C) was determined from 96 values, i.e., from 24 values for one concentration of the UV-additive multiplied by 4 concentrations.

3.2. Adhesion Evaluated at Outdoor Weathering

The second experiment with the un-weathered and also with the 14 week, 28 week, and 42 week outdoor weathered coated spruce samples obtained other findings related to the adhesion strength.

The time prolongation (τ) of the outdoor weathering from 0 to 42 weeks caused in all cases (i.e., for three modes of fungicidal pre-treatment of wood; for two modes of plasma modification of wood; for two coating types) a significant decrease in the adhesion strength between the coatings and the wood surfaces, in summary from 3.01–2.12 MPa to 1.81–1.20 MPa (Tables 2–4, Figure 3). This knowledge was confirmed by the exponential correlations "Adhesion = a + b × exp (k × τ)" with high values of the coefficient of determination R^2: (a) taking into account the different fungicidal pre-treatments of wood, the R^2 ranged from 0.903 to 0.997 (Tables 2–4), and (b) without taking into account the different fungicidal pre-treatments of wood, i.e., at the summary evaluation of the experiments when only the effects of coating type "acrylic or alkyd" and the plasma mode of wood modification "without or with" were taken into account, the R^2 ranged from 0.827 to 0.991 (Figure 3).

Table 2. The significant negative effects of outdoor weathering on the adhesion strength between the coating (acrylic or alkyd) and the native Norway spruce wood—subjected to or not subjected to plasma modification.

Weathering τ (Weeks)	Adhesion (MPa)			
	Native—Plasma Wood		Native—Natural Wood	
	Acrylic	Alkyd	Acrylic	Alkyd
0	2.82 (0.58)	2.75 (0.33)	2.20 (0.51)	2.58 (0.31)
14	1.87 (0.37)	2.18 (0.50)	1.64 (0.65)	1.99 (0.38)
28	1.37 (0.39)	1.53 (0.38)	1.44 (0.40)	1.84 (0.22)
42	1.31 (0.21)	1.51 (0.42)	1.20 (0.22)	1.42 (0.29)
"Adhesion = a + b × exp (k×τ)"				
a	1.165	1.119	1.019	0.689
b	1.661	1.653	1.173	1.866
k	−0.065	−0.039	−0.041	−0.021
R^2	0.995	0.965	0.991	0.966

Notes: the mean value is from 16 measurements (2 UV-additives "HALS or BTZ"; 4 concentrations of each UV-additive "0, 0.25, 0.5 and 1 wt.%"; 2 testing places on replicate). This mode of evaluation could be used on the basis of the knowledge that the adhesion strength was not significantly influenced by the UV-additive, i.e., by its type and concentration in the coating (see Section 3.1; Table 1 and Figure 2). The standard deviation is in the parentheses. A significant drop in the adhesion at the prolonged time of weathering (τ) was evaluated by the exponential equation "Adhesion = a + b × exp (k × τ)" from 64 values, with R^2 always above 0.95.

Table 3. The significant negative effect of outdoor weathering on the adhesion between the coating (acrylic or alkyd) and the Norway spruce wood pre-treated with fungicide H_3BO_3 —subjected to or not subjected to plasma modification.

Weathering τ (Weeks)	Adhesion (MPa)			
	H_3BO_3—Plasma Wood		H_3BO_3—Natural Wood	
	Acrylic	Alkyd	Acrylic	Alkyd
0	2.90 (0.44)	2.77 (0.34)	2.12 (0.43)	2.36 (0.42)
14	1.77 (0.47)	2.29 (0.38)	1.45 (0.46)	1.75 (0.23)
28	1.32 (0.24)	1.87 (0.23)	1.40 (0.14)	1.56 (0.28)
42	1.30 (0.28)	1.74 (0.19)	1.24 (0.13)	1.55 (0.14)
"Adhesion = a + b × exp (k×τ)"				
A	1.211	1.374	1.279	1.522
B	1.692	1.405	0.838	0.839
K	−0.083	−0.034	−0.103	−0.095
R^2	0.997	0.993	0.980	0.998

Notes: the mean value is from 16 measurements (see the first note in Table 2). The standard deviation is in the parentheses. A significant drop in the adhesion at the prolonged time of weathering (τ) was evaluated by the exponential equation "Adhesion = a + b × exp (k × τ)" from 64 values, with R^2 always above 0.95.

Another basic finding was obtained for the plasma modified spruce samples in the un-weathered state. Their adhesion strength with the acrylic and alkyd coatings was higher in comparison to the natural (plasma un-modified) spruce samples: i.e., the adhesion strength of the plasma modified wood was on average higher by 16.5% in testing the native chemically un-pre-treated wood (2.785 to 2.39 MPa), by 26.6% in testing the H_3BO_3 pre-treated wood (2.835 to 2.24 MPa) and by 13.6% in testing the BAC pre-treated wood (2.845 to 2.505 MPa)—(Tables 2–4). The plasma modification of the wood surfaces had a more positive effect on the adhesion strength of the acrylic coatings, increased by 20.5% (from 6.5 to 7.83 MPa), than the alkyd coatings, increased by 5.5% (from 7.77 to 8.2 MPa). This result can be explained by the better wettability of the hydrophilic plasma modified spruce surfaces with more polar water-borne acrylic coatings (see Section 3.3). However, as a result of the outdoor weathering, the positive effect of plasma on the adhesion strength disappeared in most cases (Tables 2–4), which was more apparent for samples painted with acrylic coatings, which manifested less durable to sunlight, water and other weathering agents (Figure 3).

Table 4. The significant negative effect of the outdoor weathering on the adhesion between the coating (acrylic or alkyd) and the Norway spruce wood pre-treated with fungicide BAC —subjected to or not subjected to plasma modification.

Weathering τ (Weeks)	Adhesion (MPa)			
	BAC—Plasma Wood		BAC—Natural Wood	
	Acrylic	Alkyd	Acrylic	Alkyd
0	3.01 (0.42)	2.68 (0.52)	2.18 (0.34)	2.83 (0.56)
14	1.75 (0.37)	2.23 (0.48)	1.91 (0.47)	2.43 (0.50)
28	1.28 (0.20)	2.02 (0.22)	1.50 (0.19)	1.36 (0.38)
42	1.26 (0.27)	1.81 (0.20)	1.53 (0.35)	1.37 (0.32)
"Adhesion = a + b × exp (k×τ)"				
A	1.174	1.548	1.282	0.036
B	1.839	1.126	0.916	2.860
K	−0.086	−0.034	−0.037	−0.020
R^2	0.997	0.996	0.932	0.903

Notes: the mean value is from 16 measurements (see the first note in Table 2). The standard deviation is in the parentheses. A significant drop in the adhesion at the prolonged time of weathering (τ) was evaluated by the exponential equation "Adhesion = a + b × exp (k × τ)" from 64 values, with R^2 above 0.95 or 0.90.

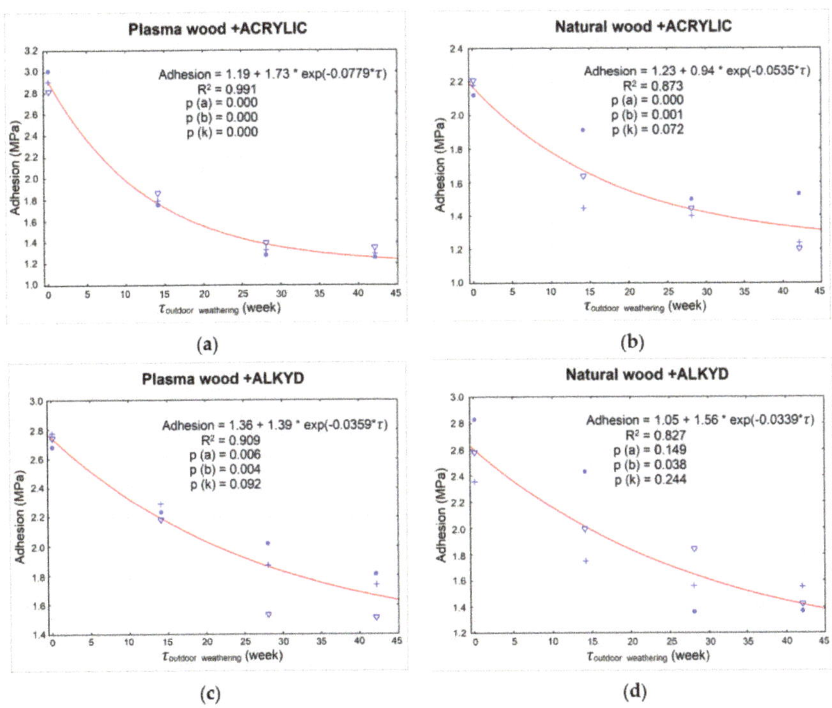

Figure 3. The summary evaluation—negative effects of the outdoor weathering prolongation (from 0 to 42 weeks) on the adhesion strength between the acrylic (**a**,**b**) or alkyd (**c**,**d**) coatings and the plasma modified spruce wood (**a**,**c**) or the natural spruce wood (**b**,**d**). Notes: each point in the graphs represents the mean value of the adhesion from 16 measurements—performed on 2 places of 8 different replicate types (2 UV-additives used in 4 concentrations). The coefficients of determination R^2 for the exponential equation "Adhesion = a + b × exp (k × τ)" (evaluating 12 mean values, i.e., 3 fungicidal pre-treatments: ∇ native wood + H_3BO_3 pre-treated wood, • BAC pre-treated wood, multiplied by the 4 outdoor weathering times) were high, from 0.827 to 0.991—confirming they have significance.

The presence of fungicides H_3BO_3 and BAC in spruce samples did not affect the adhesion strength of acrylic and alkyd coatings to their surfaces. In the complex evaluation—taking into account two coating types, two modes of wood modification with plasma, and the four outdoor weathering times—the average adhesion strength was comparable for the native wood (1.853 MPa), the wood pre-treated with H_3BO_3 (1.837 MPa), and the wood pre-treated with BAC (1.947 MPa) (averages calculated from Tables 2–4).

Comprehensively, comparing the adhesion strength of acrylic and alkyd coatings to spruce samples—i.e., for the four outdoor weathering times (0, 14, 28, 42 weeks), taking into account three modes of fungicidal pre-treatment of wood and also two modes of plasma modification of wood surfaces—the adhesion strength of acrylic coatings was partly lower (1.74 MPa) than alkyd coatings (2.016 MPa) (averages calculated from Tables 2–4). The adhesion strength of tested coating types comparably worsened with the time prolongation of the outdoor weathering. For the coated natural (plasma un-modified) spruce samples, the adhesion decreased less using acrylic coatings (by 38.92%, from 2.166 to 1.323 MPa) than using alkyd coatings (by 44.13%, from 2.59 to 1.447 MPa). On the contrary, for the coated plasma modified wood, the adhesion strength when using the acrylic coatings decreased more (by 55.67%, from 2.91 to 1.29 MPa) than when using the alkyd coatings (by 38.22%, from 2.73 to 1.687 MPa)—(Figure 3, and Tables 2–4).

3.3. *Opinions of Some Researchers on the Adhesion of Coatings to Wood Surfaces under the Influence of Modification and Weathering Impacts*

Several studies have reported that the adhesion strength between an individual coating type and an individual wood species depends first of all on the chemical and physical characteristics of the contacting materials and the aging conditions of coated wood products.

Moya et al. [14] found that the addition of TiO_2 nanoparticles, used as an UV-additive, to a water-based coating in the amount of 1% or 1.5% did not significantly affect the adhesion strength of the modified coating to seven (from nine tested) tropical wood species. This result is in line with our results concerning the un-significant impact of HALS and BTZ UV-additives in coatings on the adhesion strength (see Section 3.1).

Rehn et al. [39], Acda et al. [41] and Wolkenhauer et al. [55] found that the initial activation of wood surfaces with the atmospheric pressure plasma improved the paint adhesion and bondability by up to 30%. Their conclusions are in accordance with our results related to the water-borne acrylic coatings (see Section 3.2).

Riedl et al. [51] demonstrated that with prolonging plasma modification of maple sugar wood, its wettability (dispersive adhesion) improved more than its adhesion strength with the UV-cured polyurethane/polyacrylate coating. This is because, in addition to the free surface energy (dispersive adhesion), a number of other phenomena drive the adhesion dynamics, including surface roughness (mechanical adhesion), the formation of chemical bonds (chemical adhesion), presence of electrical discharges (electrostatic adhesion), and solubility of the coating (diffusive adhesion). Jablonský et al. [45] recommend applying water-borne coating systems on wood surfaces right after the plasma treatment, when the activated wood surface with a higher portion of polar functional groups adopts the highest degree of hydrophilicity. This is in accordance with experiments performed by Král et al. [47], where the X-ray photoelectron spectroscopy study confirmed that in beech wood surfaces were created polar functional groups at exposure to atmospheric pressure plasma, which have a positive effect on the wettability.

Avramidis et al. [56] found a positive effect of plasma on the adhesion between polyvinyl acetate (PVAc) glue and beech wood pre-treated with synthetic waxes or montan-ester waxes, because a force in the sensitive peel test increased by about 50–60%. This result is not entirely in agreement with the results of our experiment, where the plasma modification of spruce wood pre-treated with fungicide not unequivocally increased the adhesion of acrylic and alkyd coatings (see Tables 2–4). In the un-weathered state of the coated samples, the combined effect of fungicide and plasma on the adhesion strength was positive with the application of water-borne acrylic coatings (increased by 2.8% for boric acid, or by 6.7% for BAC); however, with the application of alkyd coatings, it was either negligibly positive (increased by 0.7% for boric acid) or even negative (drop by 2.55% for BAC). The different results achieved by us and by [56] can be explained by the non-polar characteristics of waxes and vice versa by the high polarity of fungicides used in our work; however, some other factors could play a role as well.

Moghaddam et al. [49] also searched the combined "chemical substance and plasma" effect on the wettability characteristics of wood surfaces. They created a thin transparent superhydrophobic layer on pine veneer surfaces from TiO_2 with the liquid flame spray technique and following the deposition of plasma polymerized perfluoro-hexane or hexamethyldisiloxane. The hydrophobic effect of this approach was stable even after repeated wetting cycles and was higher compared to that of plasma modified surfaces.

4. Conclusions

- The HALS and BTZ UV-additives applied to coatings in concentrations from 0 to 1 wt.% had no significant positive or negative effects on the adhesion strength between the acrylic or alkyd coatings and the Norway spruce wood.
- Before the outdoor weathering, alkyd coatings had a slightly better adhesion strength to spruce surfaces by 4.7% (in summary 2.66 MPa: to natural wood 2.59 MPa and to plasma modified wood 2.73 MPa) than acrylic coatings (in summary 2.54 MPa: to natural wood 2.17 MPa and to plasma modified wood 2.91 MPa). After 42 weeks of outdoor weathering, the adhesion strength of alkyd coatings in comparison to acrylic coatings was better by 19.8%, as in summary the adhesion strength of alkyd coatings dropped to 1.57 MPa while for acrylic coatings the adhesion strength dropped to 1.31 MPa. Therefore, the acrylic coatings were less durable to sunlight, water and other weathering agents than the alkyd ones.
- The fungicidal pre-treatment of spruce samples with boric acid (H_3BO_3) or benzalkonium chloride (BAC) did not affect the adhesion of coatings to wood surfaces. At a summary evaluation, i.e., taking into account two coating types "acrylic and alkyd", two modes of plasma modification of wood surfaces "natural wood and plasma wood", and four outdoor weathering times "0, 14, 28 and 42 weeks", the average adhesion strength between wood surfaces and coatings was comparable for the native wood (1.85 MPa), the wood pre-treated with H_3BO_3 (1.84 MPa), and the wood pre-treated with BAC (1.95 MPa).
- In the un-weathered state, the plasma modified wood had better adhesion with the coatings than the natural wood, with apparently more adhesion with acrylic coatings (increased by 20.5%, from 6.5 to 7.83 MPa) than with alkyd coatings (increased by only 5.5%, from 7.77 to 8.2 MPa). However, after the outdoor weathering of the coated spruce samples, the positive effect of plasma disappeared, mainly for the acrylic coatings (see point two in conclusions).
- Prolongation of the outdoor weathering from 0 to 42 weeks was associated with a significant negative exponential impact on the adhesion strength between used coatings and spruce surfaces, at which it, in summary, decreased from 3.01–2.12 MPa to 1.81–1.20 MPa.

Author Contributions: Conceptualization, L.R.; methodology, L.R.; software, L.R. and M.Š.; validation, L.R. and R.T.; formal analysis, L.R. and M.Š.; investigation, L.R., R.T. and M.Š.; resources, L.R. and M.Š.; data curation, L.R. and M.Š.; writing—original draft preparation, L.R.; writing—review and editing, L.R. and R.T.; visualization, L.R. and M.Š.; supervision, L.R.; project administration, L.R.; funding acquisition, L.R. All authors have read and agreed to the published version of the manuscript.

Funding: This work was supported by the Slovak Research and Development Agency under contract no. APVV-17-0583.

Acknowledgments: We would like to thank the company Chemolak Smolenice Ltd., Slovakia for providing part of the materials for this research.

Conflicts of Interest: The authors declare no conflict of interest.

References

1. Kataoka, Y.; Kiguchi, M.; Williams, R.S.; Evans, P.D. Violet light causes photodegradation of wood beyond the zone affected by ultraviolet radiation. *Holzforschung* **2007**, *61*, 23–27. [CrossRef]
2. Reinprecht, L.; Vidholdová, Z.; Iždinský, J. Bacterial and mold resistance of selected tropical wood species. *BioResources* **2020**, *15*, 5198–5209.
3. Aloui, F.; Ahajii, A.; Irmouili, Y.; Goerge, B.; Charrier, B.; Merlin, A. Inorganic UV absorbers for the photostabilisation of wood-clearcoating systems: Comparison with organic UV absorbers. *Appl. Surf. Sci.* **2007**, *253*, 3737–3745. [CrossRef]
4. Reinprecht, L.; Baculák, J.; Pánek, M. Natural and accelerated ageing of pains for wooden windows. *Acta Fac. Xylologiae Zvolen* **2011**, *53*, 21–31.
5. Forsthuber, B.; Schaller, C.; Grüll, G. Evaluation of the photo stabilizing efficiency of clear coatings comprising organic UV absorbers and mineral UV screeners on wood surfaces. *Wood Sci. Technol.* **2013**, *47*, 281–297. [CrossRef]
6. Poubel, D.D.; Garcia, R.A.; Lelis, R.C.C.; Riedl, B. Effect of ZnO nanoparticles on UV resistance of the heat-treated pine wood. *Sci. For.* **2017**, *45*, 49–62.
7. Pánek, M.; Oberhofnerová, E.; Zeidler, A.; Šedivka, P. Efficacy of hydrophobic coatings in protecting oak wood surfaces during accelerated weathering. *Coatings* **2017**, *7*, 172. [CrossRef]
8. Iždinský, J.; Reinprecht, L.; Nosál, E. Antibacterial efficiency of silver and zinc-oxide nanoparticles in acrylate coating for surface treatment of wooden composites. *Wood Res.* **2018**, *63*, 365–372.
9. Pánek, M.; Reinprecht, L. Colour stability and surface defects of naturally aged wood treated with transparent paints for exterior constructions. *Wood Res.* **2014**, *59*, 421–430.
10. Ozgenc, O.; Hiziroglu, S.; Yildizc, U.C. Weathering properties of wood species treated with different coating applications. *BioResources* **2012**, *7*, 4875–4888. [CrossRef]
11. Teacă, C.A.; Roşu, D.; Bodîrlău, R.; Roşu, L. Structural changes in wood under artificial UV light irradiation determined by FTIR spectroscopy and color measurements—A brief review. *BioResources* **2013**, *8*, 1478–1507. [CrossRef]
12. Cristea, M.V.; Riedl, B.; Blanchet, P. Enhancing the performance of exterior waterborne coatings for wood by inorganic nanosized UV absorbers. *Prog. Organ. Coat.* **2010**, *69*, 432–441. [CrossRef]
13. Blanchard, V.; Blanchet, P. Color stability for wood products during use: Effects of inorganic nanoparticles. *BioResources* **2011**, *6*, 1219–1229.
14. Moya, R.; Rodríguez-Zúñiga, A.; Vega-Baudrit, J.; Puente-Urbina, A. Effects of adding TiO2 nanoparticles to a water-based varnish for wood applied to nine tropical woods of Costa Rica exposed to natural and accelerated weathering. *J. Coat. Technol. Res.* **2017**, *14*, 141–152. [CrossRef]
15. Schaller, C.; Rogez, D. New approaches in wood coating stabilization. *J. Coat. Technol. Res.* **2007**, *4*, 401–409. [CrossRef]
16. Forsthuber, B.; Grüll, G. The effect of HALS in the prevention of photo-degradation of acrylic clear topcoats and wooden surfaces. *Polym. Degrad. Stabil.* **2010**, *95*, 746–755. [CrossRef]
17. Samyn, P.; Stanssens, D.; Paredes, A.; Becker, G. Performance of organic nanoparticle coatings for hydrophobization of hardwood surfaces. *J. Coat. Technol. Res.* **2014**, *11*, 461–471. [CrossRef]
18. Meijer, M.D. Review on the durability of exterior wood coatings with reduced VOC-content. *Prog. Organ. Coat.* **2001**, *43*, 217–225. [CrossRef]
19. Grüll, G.; Tscherne, F.; Spitaler, I.; Forsthuber, B. Comparison of wood coating durability in natural weathering and artificial weathering using fluorescent UV-lamps and water. *Eur. J. Wood Wood Prod.* **2014**, *72*, 367–376. [CrossRef]
20. Reinprecht, L. Fungicides for Wood Protection—World Viewpoint and Evaluation/Testing in Slovakia. In *Fungicides*; Carisse, O., Ed.; InTech: Rijeka, Croatia, 2010; pp. 95–122.
21. Weththimuni, M.L.; Capsoni, D.; Malagodi, M.; Licchelli, L. Improving wood resistance to decay by nanostructured ZnO-based treatments. *J. Nanomater.* **2019**, *2019*, 6715756. [CrossRef]
22. Nejad, M. Coating Performance on Preservative Treated Wood. Ph.D. Thesis, Department of Forestry, University of Toronto, Toronto, ON, Canada, 2011; p. 154.
23. Daniels, T.; Hirsch, M.; McClelland, K.; Ross, A.; Williams, R.S. Clear exterior finishes: Finding the balance between aesthetics and durability. *JCT Coat. Tech.* **2004**, *9*, 42–48.

24. Williams, R.S.; Jourdain, C.; Daisey, G.; Springate, R.W. Wood properties affecting finish service life. *J. Coat. Technol.* **2000**, *72*, 35–42. [CrossRef]
25. Wagenführ, R. *Holzatlas*, 6th ed.; Fachbuchverlag: Leipzig, Germany, 2007; p. 816.
26. Tascioglu, C.; Yalcin, M.; Sen, S.; Akcay, C. Antifungal properties of some plant extracts used as wood preservatives. *Int. Biodeterior. Biodegrad.* **2013**, *85*, 23–28. [CrossRef]
27. Kúdela, J. Wetting of wood surface by a liquids of a different polarity. *Wood Res.* **2014**, *59*, 11–24.
28. Jankowska, A.; Boruszewski, P.; Drozdzek, M.; Rebkowski, B.; Kaczmarczyk, A.; Skowronska, A. The role of extractives and wood anatomy in the wettability and free surface energy of hardwoods. *BioResources* **2018**, *13*, 3082–3097. [CrossRef]
29. Zhang, H.; Kai, T.U.; Hou, Q.; Lin, H.; Quan, L.I. The physiological and biochemical mechanisms of *Cinnamomum camphora* xylem extracts inhibit wood-decay fungi. *Wood Res.* **2020**, *65*, 531–542. [CrossRef]
30. Reinprecht, L. *Wood Deterioration, Protection and Maintenance*, 1st ed.; John Wiley & Sons Ltd.: Chichester, UK, 2016; p. 357.
31. Cheng, E.; Sun, X. Effects of wood-surface roughness, adhesive viscosity and processing pressure on adhesion strength of protein adhesive. *J. Adhes. Sci. Technol.* **2006**, *20*, 997–1017. [CrossRef]
32. Kaygin, B.; Akgun, E. Comparison of conventional varnishes with nanolacke UV varnish with respect to hardness and adhesion durability. *Int. J. Mol. Sci.* **2008**, *9*, 476–485. [CrossRef]
33. Kúdela, J.; Liptáková, E. Adhesion of coating materials to wood. *J. Adhes. Sci. Technol.* **2012**, *20*, 875–895. [CrossRef]
34. Odrášková, M.; Rahel, J.; Zahoranová, A.; Tiňo, R.; Černák, M. Plasma activation of the wood surface by diffuse coplanar surface barrier discharge. *Plasma Chem. Plasma Process.* **2008**, *28*, 203–211. [CrossRef]
35. Viöl, W. Possibilities of cold plasma treatment at atmospheric pressure to modify wood surfaces and relevant applications. In Proceedings of the 18th International Conference on Surface Modification of Materials by Ion Beams, Wood and Organic Materials, Kusadasi, Turkey, 15–20 September 2013; Volume 9, p. 2013.
36. Nguyen, T.; Chen, W.; Cao, Y.; Wang, X.; Shi, S.; Chen, M.; Zhou, X.; Nguyen, Q. Improving bonding strength of oven-dried poplar veneers using atmospheric cold plasma treatment. *BioResources* **2018**, *13*, 1843–1851. [CrossRef]
37. Žigon, J.; Petrič, M.; Dahle, S. Dielectric barrier discharge (DBD) plasma treatment of lignocellulosic materials in air at atmospheric pressure for their improved wettability: A literature review. *Holzforschung* **2018**, *72*, 979–991. [CrossRef]
38. Hünnekens, B.; Avramidis, G.; Ohms, G.; Krause, A.; Viöl, W.; Militz, H. Impact of plasma treatment under atmospheric pressure on surface chemistry and surface morphology of extruded and injection-molded wood-polymer composites (WPC). *Appl. Surf. Sci.* **2018**, *441*, 564–574. [CrossRef]
39. Rehn, P.; Wolkenhauer, A.; Bente, M.; Förster, S.; Viöl, W. Wood surface modification in dielectric barrier discharges at atmospheric pressure. *Surf. Coat. Technol.* **2003**, *174–175*, 515–518. [CrossRef]
40. Wolkenhauer, A.; Avramidis, G.; Hauswald, E.; Militz, H.; Viöl, W. Sanding vs. plasma treatment of aged wood: A comparison with respect to surface energy. *Int. J. Adhes. Adhes.* **2009**, *29*, 18–22. [CrossRef]
41. Acda, M.N.; Devera, E.E.; Cabangon, R.J.; Ramos, H.J. Effects of plasma modification on adhesion properties of wood. *Int. J. Adhes. Adhes.* **2012**, *32*, 70–75. [CrossRef]
42. Avramidis, G.; Klarhöfer, L.; Maus-Friedrischs, W.; Militz, H.; Viöl, W. Influence of air plasma treatment at atmospheric pressure on wood extractives. *Polym. Degrad. Stabil.* **2012**, *97*, 469–471. [CrossRef]
43. Novák, I.; Chodák, I.; Sedliačik, J.; Vanko, V.; Matyašovský, J.; Šivová, M. Pre-treatment of beech wood by cold plasma. *Forest. Wood Technol.* **2013**, *83*, 288–291.
44. Reinprecht, L.; Tiňo, R.; Šomšák, M. Adhesion of coatings to plasma modified wood at accelerated weathering. In Proceedings of the European Conference on Wood Modification 2018 (ECWM9), Arnhem, The Netherlands, 17–18 September 2018; p. 54.
45. Jablonský, M.; Šmatko, L.; Botková, M.; Tino, R.; Šima, J. Modification of wood wettability (European Beech) by diffuse coplanar surface barrier discharge plasma. *Cellul. Chem. Technol.* **2014**, *50*, 41–48.
46. Potočňáková, L.; Hnilica, J.; Kudrle, V. Increase of wettability of soft- and hardwoods using microwave plasma. *Int. J. Adhes. Adhes.* **2013**, *45*, 125–131. [CrossRef]
47. Král, P.; Ráhel, J.; Stupavská, M.; Šrajer, J.; Klímek, P.; Mishra, K.P.; Wimmer, R. XPS depth profile of plasma–activated surface of beech wood (*Fagus sylvatica*) and its impact on polyvinyl acetate tensile shear bond strength. *Wood Sci. Technol.* **2015**, *49*, 319–330. [CrossRef]

48. Yousoo, H.; Manolach, S.O.; Denes, F.; Rowell, R.M. Cold plasma treatment on starch foam reinforced with wood fiber for its surface hydrophobicity. *Carbohydr. Polym.* **2011**, *86*, 1031–1037.
49. Moghaddam, S.M.; Heydari, G.; Tuominen, M.; Fielden, M.; Haapanen, J.; Mäkelä, M.J.; Wålinder, E.P.M.; Claesson, M.P.; Swerin, A. Hydrophobisation of wood surfaces by combining liquid flame spray (LFS) and plasma treatment: Dynamic wetting properties. *Holzforschung* **2016**, *70*, 527–537. [CrossRef]
50. EN ISO 4287. *Geometrical Product Specifications (GPS). Surface Texture: Profile Method—Terms, Definitions and Surface Texture Parameters*; European Committee for Standardization: Brussels, Belgium, 1997.
51. Riedl, B.; Angel, C.; Prégent, J.; Blanchet, P.; Stafford, L. Wood surface modification by atmospheric-pressure plasma and effect on waterborne coating adhesion. *LignoCellulose* **2013**, *2*, 292–306. [CrossRef]
52. EN 927-6. *Paints and Varnishes. Coating Materials and Coating Systems for Exterior Wood. Part 6: Exposure of Wood Coatings to Artificial Weathering Using Fluorescent UV Lamps and Water*; European Committee for Standardization: Brussels, Belgium, 2018.
53. EN 927-3. *Paints and Varnishes. Coating Materials and Coating Systems for Exterior Wood. Part 3: Natural Weathering Test*; European Committee for Standardization: Brussels, Belgium, 2012.
54. EN ISO 4624. *Paints and Varnishes. Pull-off Test for Adhesion*; European Committee for Standardization: Brussels, Belgium, 2016.
55. Wolkenhauer, A.; Avramidis, G.; Hauswald, E.; Militz, H.; Viöl, W. Plasma treatment of wood-plastic composites to enhance their adhesion properties. *J. Adhes. Sci. Technol.* **2008**, *22*, 2025–2037. [CrossRef]
56. Avramidis, G.; Scholz, G.; Nothnick, E.; Militz, H.; Viöl, W.; Wolkenhauer, A. Improved bondability of wax-treated wood following plasma treatment. *Wood Sci. Technol.* **2011**, *45*, 359–368. [CrossRef]

coatings

MDPI

Article

Caffeine and TiO_2 Nanoparticles Treatment of Spruce and Beech Wood for Increasing Transparent Coating Resistance against UV-Radiation and Mould Attacks

Miloš Pánek [1,*], Kristýna Šimůnková [1], David Novák [1], Ondřej Dvořák [1], Ondřej Schönfelder [1], Přemysl Šedivka [1] and Klára Kobetičová [2]

[1] Department of Wood Processing and Biomaterials, Faculty of Forestry and Wood Sciences, Czech University of Life Sciences, Kamýcká 129, 165 00 Prague, Czech Republic; simunkovak@fld.czu.cz (K.Š.); novakd@fld.czu.cz (D.N.); dvorak18@fld.czu.cz (O.D.); schonfelder@fld.czu.cz (O.S.); sedivka@fld.czu.cz (P.Š.)

[2] Department of Materials Engineering and Chemistry, Faculty of Civil Engineering, Czech Technical University in Prague, 166 29 Prague, Czech Republic; klara.kobeticova@fsv.cvut.cz

* Correspondence: panekmilos@fld.czu.cz; Tel.: +420-224-383-867

Received: 10 November 2020; Accepted: 21 November 2020; Published: 24 November 2020

Abstract: The effect of the initial modification of beech and spruce wood using a solution of caffeine and of a commercial product FN-NANO® FN-1 containing a water dispersion of TiO_2 nanoparticles for increasing the service life of a transparent oil and acrylate coatings during 6 weeks of artificial accelerated weathering was tested. Changes in colour, gloss, and the contact angle of water were monitored. Degradation of the coating film was also evaluated visually and microscopically. The resistance of the coatings to mould growth was also subsequently tested. Based on the results, it is possible to recommend the initial treatment of spruce and beech wood with a 2% caffeine solution or 15% solution of FN-NANO® dispersion to increase the overall life of a transparent acrylic coating in exterior applications. No positive effect of the applied treatments was observed with the oil coating. In addition, lower concentrations of FN-NANO® did not have a sufficient effect, and the synergistic effect of using FN-NANO® in a mixture with a 1% caffeine solution was also not confirmed.

Keywords: wood; caffeine; TiO_2 nanoparticles; transparent coatings; UV-resistance; mould attack

1. Introduction

Wood is a material of natural origin which is subject to a relatively rapid loss of its original appearance in exterior applications [1]. This is caused both by abiotic influences [2] and biotic infestation, in particular by moulds and wood-staining fungi [3,4]. The original appearance of wood in an exterior location can be further preserved using coating systems [5]. Transparent coatings can preserve not only the original design but also partly the colour of the base wood; however, their long-term stability is reduced compared to pigmented coatings by deeper and more intense penetration of sunlight into the coating film and base wood [6,7]. Another factor causing faster defoliation of coatings is the growth of fungal hyphae on their surface and in the zone between the coating and the wood surface [8,9]. In practice, penetrating base coatings containing fungicides and photo-stabilising components are most often used to suppress fungal growth, increase the bio-resistance of non-durable wood species, and increase photostability [5,10]. Ultraviolet (UV) stabilisers, hindered amine light stabilisers (HALS), and nanoparticles [1–13] are most often used as additives, and substances based on triazoles, carbamates, and others are most often used as fungicides [14,15].

An interesting possibility for increasing bio-resistance in parallel with photo-stabilisation of wood is the use of TiO_2 nanoparticles in anatase form [2,13,16–18]. The use of various substances of natural origin extractable from renewable sources is a more ecologically acceptable form of wood protection against bio-attack [19]. One of the substances with a proven biocidal effect is caffeine [20–22], but it is also disadvantageous in terms of its washability from wood [23]. However, this disadvantage would be eliminated through use of the top barrier layer of the coating system, which protects the base caffeine-impregnated wood. Nevertheless, it is necessary to determine whether the initial pre-treatment with TiO_2 nanoparticles and caffeine adversely affects the overall service life of the applied top coating, as described in some works [24,25]. Woods of the Norway spruce (*Picea abies* L. Karst) and European beech (*Fagus sylvatica* L.) were selected for research in this work. Spruce wood is widely used in exterior structures without a shelter—class 3 by EN 335 [26]. This is a type of wood with lower resistance to bio-attack [27] and its additional protection against bio-attack is often required. Transparent exterior coatings on this type of wood do not have optimal durability [28,29]. Beech wood is rarely used outdoors due to its low resistance to bio-attacks [27]. Its deep impregnation is necessary, and creosote oils are used, for example, for its use on railway sleepers [5]. Additionally, due to climate change, it is possible to anticipate an increase in its share in the total volume of wood processed—here, in Central Europe in particular [3,30,31]. The use of a suitable coating system and fungicidal components represents one of the simplest and cheapest variants of its possible use in class 3 by EN 335 [26]—outdoors without a shelter and without contact with the ground and water.

The aim of the experiment was to determine how the overall service life of the acrylate and oil coating system is affected by the initial surface modification of the underlying spruce and beech wood using caffeine solutions, dispersion of TiO_2 nanoparticles, and their combinations. The method of artificial accelerated weathering in a UV-chamber and a subsequent test of mould growth on both ageless and weathered surfaces of treated wood were used for testing.

2. Material and Methods

2.1. Wood Samples

Norway spruce (*Picea abies*, L. Karst) wood with an average density of 412 $kg \cdot m^{-3}$ and beech wood (*Fagus sylvatica*, L.) with an average density of 710 $kg \cdot m^{-3}$ were used for the experiments. Test specimens measuring 20 mm × 40 mm × 150 mm with a milled test area of 40 mm × 150 mm were prepared. Prior to treatment, the samples were conditioned in the laboratory at $T = 20$ °C and in air humidity of $\phi = 65\%$ to an equilibrium humidity of 12%, and they were subsequently modified for further testing.

2.2. Treatments of Samples

In the first step, the test specimens were soaked in more concentrations of impregnation solutions containing an aqueous dispersion of commercial product FN-NANO© (FN-1) (FN-NANO® s.r.o., Prague, Czech Republic). FN-NANO® dispersion containing about 7.5% TiO_2 nanoparticles in anatase form with the addition of 2.4% $ZnSO_4$ fulfilling the function of binder. Other sets were impregnated in 2% caffeine solution (Sigma-Aldrich, Prague, Czech Republic) and sets with different concentrations of FN-NANO® in solution combined with 1% of caffeine were also prepared (see Table 1). The intake of the solution and the soaking time were adjusted and controlled by continuous weighing of the samples so that the intake of impregnating substance into the wood was about 120 ± 20 $kg \cdot m^{-3}$ for both types of wood (beech is significantly more permeable compared to spruce). Subsequently, the samples were dried again in the laboratory at $T = 20$ °C and in air humidity of $\phi = 65\%$ to an equilibrium humidity of 12%. After drying, in the second step, top coatings were applied by brush to the samples. Acrylic (AC) transparent exterior glaze (Impranal profi with UV-filter, Stachema a.s., Kolín, Czech Republic) and transparent coating based on vegetable oils (OL) for exterior OSMO UV 420 (Osmo, Münster, Germany). Both were applied in two layers by the producers' recommendations with a deposit of approximately 100 $g \cdot m^{-2}$. A total of 4 test specimens were prepared for each type of treatment (see Table 1) and testing. After drying the coating systems, tests of artificial accelerated weathering were performed.

Table 1. Tested sets of specimens and their specification.

Types of Samples	Solution of FN-NANO® Dispersion (Concentration)	Caffeine Solution (Concentration)	Acrylic Coating (AC)	Oil-Based Coating (OL)
R-R	–	–	–	–
R-A	–	–	2 layers	–
1-A	10%	–	2 layers	–
2-A	15%	–	2 layers	–
3-A	10%	1%	2 layers	–
4-A	15%	1%	2 layers	–
5-A	–	2%	2 layers	–
R-O	–	–	–	2 layers
1-O	10%	–	–	2 layers
2-O	15%	–	–	2 layers
3-O	10%	1%	–	2 layers
4-O	15%	1%	–	2 layers
5-O	–	2%	–	2 layers

Note: The spruce samples were marked (S) and the beech samples (B). Solutions of FN-NANO® dispersion with concentrations 3.5% and 5%, as well as their combinations with a total 1% concentration of caffeine were also prepared, but the achieved results after artificial weathering were unsatisfactory, and they have therefore not been further evaluated in this article.

2.3. Artificial Accelerated Weathering

Artificial accelerated weathering (AW) was performed on the basis of EN 927-6 [32] in a UV chamber (Q-Lab, Cleveland, OH, USA) (radiation parameters 1.10 Wm^{-2}; T on black panel = 65 °C were modified) as standard with 2.5 h radiation phases and 0.5 h spraying with distilled water in the dark, and air conditioning at 45 °C once a week for 24 h. In addition, once a week, 3 temperature cycles (1 h at 80 °C and one hour at −25 °C) were inserted in the Discovery My DM340 air conditioning chamber (ACS, Massa Martana, Italy). The tests lasted for 6 weeks and the selected properties of the tested samples were evaluated at the beginning and after the 3rd and 6th week of AW.

2.4. Tested Properties—Colour, Gloss, Surface Wetting, and Visual Evaluation

The evaluated properties were colour changes measured by a spectrophotometer (CM-600d, Konica Minolta, Osaka, Japan) with an observation angle of 10°, $d/8$ geometry, D65 light source, and the SCI setting was used. The colour changes of the parameters L^* (brightness), a^* (+red; −green), b^* (+yellow, −blue), and the total colour change ΔE^* according to the known relationship $\Delta E^* = (\Delta L^2 + \Delta a^2 + \Delta b^2)^{\frac{1}{2}}$ were evaluated [33]. Twelve measurements were made for each group of test samples and AW time.

The gloss was measured with an MG268-F2 glossmeter (KSJ, Quanzhou, China) at an angle of 60° according to EN ISO 2813 [34]. Twelve measurements were performed for each group of tested samples and AW time.

Surface wetting was evaluated using Goniometer DSA 30E (Krüss, Hamburg, Germany) and Krüss software (Krüss, Hamburg, Germany). The sessile drop method was used. The dosing volume of the distilled water drop was 5 µL and measurement time of the water contact angle on the surface after deposition was 5 s. A total of 20 measurements for each type of treatment and AW time were performed.

Visual evaluation was performed using 10× folder magnification and scans of samples (using a desktop scanner at a resolution of 300 dots per inch (DPI) (Canon 2520 MFP, Canon, Tokyo, Japan) before and during AW. Microscopic analyses of surfaces used a confocal laser scanning microscope (Lext Ols 4100, Olympus, Tokyo, Japan) with 108-fold magnification.

2.5. Mould Resistance Test

The test samples without weathering and after 6 weeks of artificial accelerated weathering were tested for the resistance of the surfaces to moulds. Petri dishes with Czapek–Dox agar were left in the open for 24 h to free up mould spores. This method provides more variable results compared to the laboratory methods using pure mould cultures, but it corresponds better to the real conditions of the exposed wood. The samples were then sealed and left at 30 °C and 90% relative humidity. After checking the growth of mould in the dish, moist test specimens measuring 15 mm × 40 mm × 10 mm were placed in the dishes with growing moulds, where their upper side measuring 40 mm × 15 mm was manipulated from the surfaces of the exposed test specimens, and not exposed to artificial accelerated weathering. The samples were placed on a low plastic pad to prevent the agar from seeping into the wood. Two test specimens were placed in each dish so that one dish did not contain samples from the same set of the tested types of treatments. Mould growth was evaluated after 7, 14, 21, and 28 days of the mould test based on the ČSN 490604 [35] standard, where degree 0 corresponds to samples without growth, 0%–10% is evaluated by degree 1, 10%–30% by degree 2, 30%–50% by degree 3, and above 50% of the stand at degree 4.

2.6. Statistical Evaluation

The results of the obtained measurements were evaluated using Statistica software (version 13.4) (StatSoft, Palo Alto, CA, USA) and using mean values, standard deviations (SD), and whisker plots using means and 2 ± SD.

3. Results and Discussion

3.1. Visual and Microscopic Evaluation

Based on visual evaluation supplemented by a microscopic analysis of the surfaces of the tested samples after artificial accelerated aging, positive effect of certain types of pre-treatments improving the durability of the acrylate coating (Figure 1) were evident. Compared to untreated wood and two layers of acrylic coating (Figure 1b), these were mainly 2% caffeine solution treatments (Figure 1c), the surface treated with 15% FN-NANO® dispersion (Figure 1e), in beech wood also 10% FN-NANO® dispersions (Figure 1d), and in spruce, the combination of 1% caffeine solution and 10% FN-NANO® dispersion (Figure 1f) also had a relatively good effect. For other types of treatment, the result was better compared to the initially untreated surface, but the positive effect was not as expressive. As with other types of bio-treatments [36], the effect of caffeine treatment on wetting can be explained by waterborne acrylate coating by reducing the contact angle of water wetting and increasing surface free energy, which positively affect the adhesion of coatings to wood [37,38]. Based on our further research, it was found that the contact angle of water wetting in beech decreased after adjustment from the original 63.7° to 52.6° and from 105.6° to 83.4° in spruce, whilst surface free energy increased from 49.15 to 53.70 $mN{\cdot}m^{-2}$ in beech and from 30.5 to 43.9 $mN{\cdot}m^{-2}$ in spruce [39,40]. Regarding the positive effect of FN-NANO® containing TiO_2 nanoparticles, it was documented that at certain concentrations—in particular, if TiO_2 nanoparticles penetrate deeper into the wood—the effect on the service life of the top-coating system may not be negative, as evidenced by certain works [24,25]. In fact, they can even reduce the degradation of the under-laying wood surfaces by capturing part of the incident UV and visible radiation spectra. The potential synergistic effect of a mixture of TiO_2 nanoparticles and caffeine on improving the service life of coating systems due to weathering was not demonstrated in this work. Only in the case of spruce was the overall appearance of the coatings slightly better after the combined treatment, whereas for beech it worsened (Figure 1d,e versus Figure 1f,g). Figure 1a shows the fully degraded surface of untreated native wood with a preserved layer of non-photodegradable cellulose [41], causing significant lightening of the test specimens during artificial accelerated weathering (Figure 1a).

Figure 1. *Cont.*

(e)

(f)

(g)

Figure 1. Beech (left) and spruce (right) wood samples before (0.w.) and after accelerated weathering (AW) (6.w.). (**a**) Reference native wood (R-R); (**b**) acrylic coating on untreated wood (R-A); (**c**) acrylic coating on caffeine treated wood (5-A); (**d**) acrylic coating on wood treated with 10% FN-NANO® (1-A); (**e**) acrylic coating on wood treated with 15% FN-NANO® (2-A); (**f**) acrylic coating on wood treated with a mixture of 10% FN-NANO® and 1% of caffeine (3-A); (**g**) acrylic coating on wood treated with a mixture of 15% FN-NANO® and 1% of caffeine (4-A). Note: The real observed area of microscopic picture was 2500 μm × 2500 μm using 108-folder magnification.

Overall, the observed degradations of oil-based coating (Figure 2) were significantly higher compared to the degradations of acrylic coating (Figure 1). They were comparable for both beech and spruce wood without pre-treatment (Figure 2a,c), and the different types of treatments did not have a significant positive impact (Figure 2b,d). For this reason, colour changes, gloss changes, surface wetting, and mould growth were only subsequently evaluated for the acrylic coating system (Figures 3–5 and Table 2).

Figure 2. Oil-based coating after 6 weeks of AW. (**a**) Beech wood without pre-treatment (B-R-O); (**b**) beech wood with caffeine pre-treatment (B-5-O) achieved the best results from the tested oil-coated variants; (**c**) spruce wood without pre-treatment (S-R-O); (**d**) spruce wood with 10% FN-NANO® treatment achieved the best results from the tested oil-coated variants (S-1-O).

3.2. Colour, Gloss, and Water Contact Angle Changes during AW

During artificial accelerated weathering, there were significant colour changes in all of the tested surfaces (Figure 3). The changes were smaller for darker beech wood in cases where the coating film was preserved and there was no leaching of darker photodegraded lignin [42], and thus the overall colour change was lower than in comparable cases of overall lighter spruce wood (Figure 1). In contrast, for untreated wood, the ΔE^* was greater for beech, as only light-coloured cellulose remained in the surface zones [4]. Overall, the coated surfaces of the primary treated wood often had a higher overall colour change compared to the untreated surfaces. However, this was due to the disruption of the coating film of the untreated surfaces (R-A) (see also Figure 1) and the leaching of darker photodegraded lignins [43] and subsequent lightening, thus returning the colour to the original lighter shade of wood that had not been weathered. Of the pre-treatments that increased the overall service life of the coating film, they also had a partial effect on reducing the colour changes of 2-A for spruce and 1-A and less 2-A for beech. This confirms the positive effect of TiO_2 on the absorption and reflection of UV radiation, and thus reduces its impact on wood [13,44]. However, it was demonstrated that its effect and suitable concentrations vary for different types of wood (Figure 3). The decrease in gloss of the coatings sensitively predicts the degradation of its surface layers [45,46]. In terms of the tested treatments, a positive effect on the preservation of the original gloss was observed in caffeine-treated spruce wood (Figure 4) with an overall increased service life (Figure 1). The same was observed in beech treated with a higher concentration of TiO_2 nanoparticles B-2-A (Figure 4) with an observed positive effect on colour fastness (Figure 3), as well as the overall service life of the top acrylate coating system (Figure 1). The contact angle of wetting and the reduction of hydrophobicity of the surfaces was observed mainly in untreated photodegraded wood, particularly in beech (Figure 5). Compared to other pre-treatments, a slight decrease in coated samples was only observed for S-1-A, which also confirmed the observed effect on reducing the overall life of the coating film (Figure 1). The hydrophobic effect was otherwise preserved in the samples where the top layer of the coating film was already disturbed (Figure 1), and there were only slight differences in the variability of the observed values (Figure 5). This also corresponds to the results of the work [46], where the preservation of hydrophobicity also did not correspond to the disruption of the overall appearance of the tested oak wood samples. Overall, a relatively high variability of the achieved values was observed (Figures 3–5). This is caused by several factors. The first factor is the variability of the measured values of individual samples in the set due to the inhomogeneity of wood as a material of natural origin. The second reason is the increase in inhomogeneity of some tested surfaces due to weathering, as was documented in Figure 1. It was confirmed that the quality of the coating after aging can be evaluated overall only by a complex of evaluated characteristics, where the overall visual evaluation of the coating film must not be neglected [7].

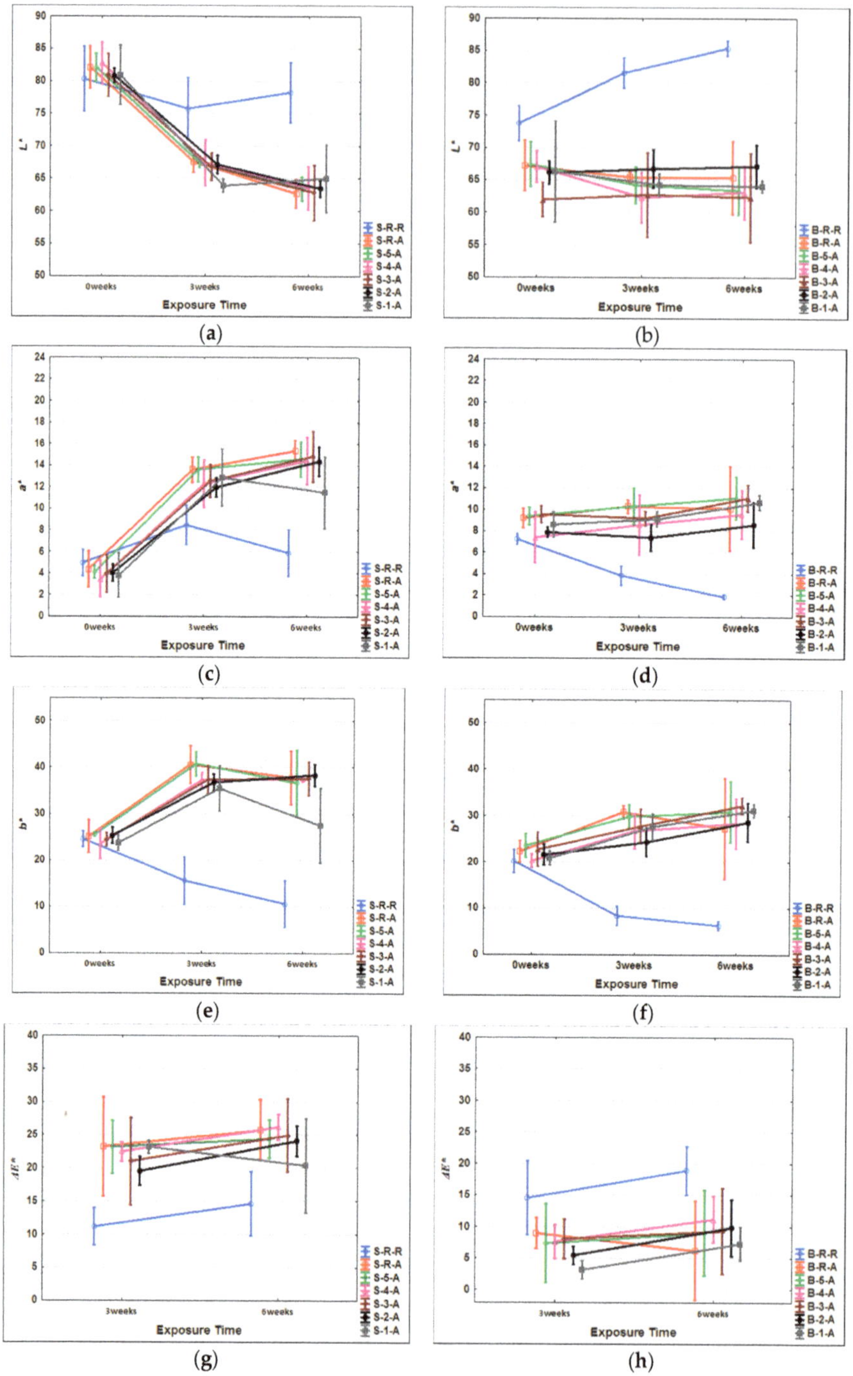

Figure 3. Colour changes in tested wooden surfaces during AW: (**a**) L^* changes on spruce; (**b**) L^* changes on beech; (**c**) a^* changes on spruce; (**d**) a^* changes on beech; (**e**) b^* changes on spruce; (**f**) b^* changes on beech; (**g**) ΔE^* changes on spruce; (**h**) ΔE^* changes on beech.

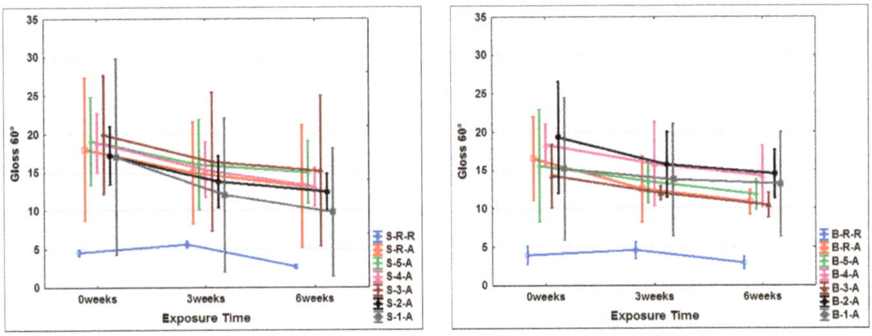

(a) (b)

Figure 4. Gloss changes in tested wooden surfaces during AW: (**a**) Gloss changes on tested spruce samples; (**b**) Gloss changes on tested beech samples.

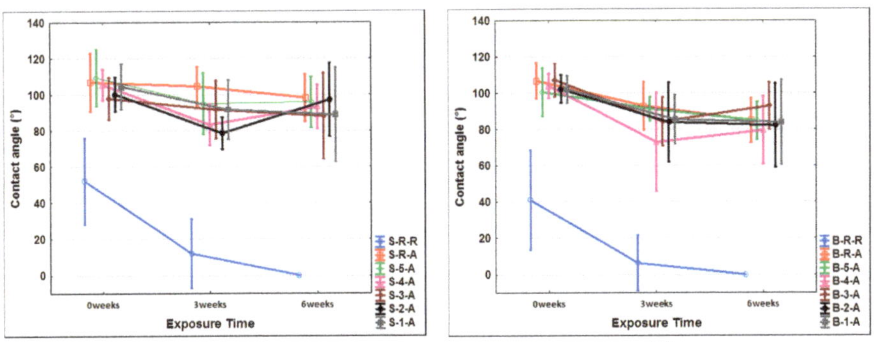

(a) (b)

Figure 5. Water contact angle changes in tested wooden surfaces during AW: (**a**) Water contact angle (°) on tested spruce samples; (**b**) Water contact angle (°) on tested beech samples.

3.3. *Mould Growth Tests*

Moulds pose a significant risk of damage to surface-treated wood products outdoors, and moulds have a significant impact on their appearance [5]. There are various methods for testing mould growth on wood and treated wood [3,4,47]. The results can vary depending on the method used [3]. One of the possibilities is to use a free fall of mould spores into a prepared nutrient medium and subsequent insertion of test specimens [48]. This method provides more variable results similar to tests in natural conditions [4]. In this work, the highest rate of observed mould growth from the tested test samples of each series was evaluated (Table 2). This increases the likelihood that the level of risk will be more accurately verified in real exposures, where different species of mould may begin to grow under appropriate conditions [47], or a combination thereof.

In general, compared to spruce, higher mould growth was observed on the surface of the tested coated beech samples (Table 2). This can be explained by the higher natural bio-resistance of spruce wood [27] and also by the better durability of the coating system after AW on spruce wood (Figure 1). Before and after AW, treatment with a 2% caffeine solution (5-A) with a biocidal effect, which was also confirmed in other works, had a positive effect on beech and spruce wood [2,23]. In comparison with the work of Kwaśniewska-Sip et al. [49], a better biocidal effect was achieved due to barrier protection of treated wood with a coating that effectively prevented caffeine leaching during 6 weeks of artificial accelerated weathering. In several cases, a positive effect of treatment with TiO_2 nanoparticles was visible [8]—in particular, at a concentration of 15% FN-NANO® (2-A and 4-A), more so in samples after AW (Table 2). This may be related to the greater exposure of the TiO_2-treated layer connected with the degradation of the acrylate coating, which in itself did not show a significant anti-mould effect. A similar result was observed for spruce wood, but the growth of mould was generally lower than for beech wood. The result confirms that only higher concentrations of TiO_2 nanoparticles have sufficient fungicidal activity, even if the specific amount required is affected by the co-modifying agent [18,50].

Table 2. Maximum mould growth on tested samples before (R) and after weathering (W).

Beech	Degree of Moulds Growth (% of Surface Covered by Moulds in Parenthesis)				Spruce	Degree of Moulds Growth (% of Surface Covered by Moulds in Parenthesis)			
	Time of Moulds Exposure					Time of Moulds Exposure			
	1 week	2 weeks	3 weeks	4 weeks		1 week	2 weeks	3 weeks	4 weeks
B-R-R-R	0 (0)	4 (80)	4 (90)	4 (90)	S-R-R-R	0 (0)	3 (50)	4 (65)	4 (60)
B-R-R-W	0 (0)	3 (50)	4 (60)	4 (75)	S-R-R-W	0 (0)	3 (50)	4 (60)	4 (90)
B-R-A-R	0 (0)	3 (50)	3 (50)	4 (75)	S-R-A-R	0 (0)	3 (40)	4 (55)	4 (65)
B-R-A-W	0 (0)	4 (75)	4 (75)	4 (75)	S-R-A-W	0 (0)	4 (75)	4 (80)	4 (95)
B-5-A-R	0 (0)	1 (5)	1 (5)	2 (10)	S-5-A-R	0 (0)	0 (0)	0 (0)	1 (5)
B-5-A-W	0 (0)	0 (0)	0 (0)	2 (25)	S-5-A-W	0 (0)	0 (0)	0 (0)	0 (0)
B-4-A-R	0 (0)	0 (0)	0 (0)	2 (25)	S-4-A-R	0 (0)	0 (0)	0 (0)	2 (10)
B-4-A-W	0 (0)	0 (0)	0 (0)	1 (5)	S-4-A-W	0 (0)	0 (0)	0 (0)	1 (1)
B-3-A-R	0 (0)	4 (75)	4 (85)	4 (100)	S-3-A-R	0 (0)	0 (0)	0 (0)	2 (10)
B-3-A-W	0 (0)	0 (0)	0 (0)	2 (25)	S-3-A-W	0 (0)	0 (0)	0 (0)	1 (1)
B-2-A-R	0 (0)	4 (75)	4 (75)	4 (100)	S-2-A-R	0 (0)	0 (0)	0 (0)	3 (35)
B-2-A-W	0 (0)	0 (0)	0 (0)	1 (5)	S-2-A-W	0 (0)	0 (0)	0 (0)	1 (1)
B-1-A-R	0 (0)	4 (75)	4 (75)	4 (100)	S-1-A-R	0 (0)	0 (0)	0 (0)	1 (5)
B-1-A-W	0 (0)	0 (0)	0 (0)	2 (25)	S-1-A-W	0 (0)	1 (5)	1 (5)	1 (5)

Note: degree 0 represents the best results; degree 4 represents the worst results.

Research also confirms that the protection of beech wood against bio-attacks by moulds outdoors using coatings is more difficult. In addition to its lower bio-resistance, its greater permeability may also have an effect [51], and thereby increased demands on the amount of applied protective substance, which does not provide sufficient film-forming protection on the wood surface. However, it is encouraging that higher concentrations of the active ingredient TiO_2 used in this research (sets 2-A and 4-A) represented a relatively effective alternative against mould growth (Table 2).

4. Conclusions

Transparent coatings on wood outdoors do not have sufficient service life and colourfastness due to more significant penetration of sunlight into the coating film and the base wood. Another significant factor causing degradation of coatings in humid environments are moulds. The possibility of increasing the durability and resistance to mould attacks in acrylic and oil coating on beech and spruce wood was investigated. Initial treatments of wood by dipping in a 2% solution of caffeine and various concentrations of commercial FN-NANO® containing TiO_2 nanoparticles and their mixtures were used. A significant positive effect of caffeine treatment on the service life of the acrylate coating system during accelerated artificial weathering was observed in both types of tested woods. The 15% FN-NANO® dispersions treatment also showed good efficiency. The combined effect of a mixture of caffeine and FN-NANO® had no significant positive effect. No improved quality of the oil coating was observed using the initial wood treatments. Degradation by moulds was significantly reduced when the wood was initially treated with a 2% caffeine solution. Higher concentrations of FN-NANO® dispersion had a positive impact—in particular, on samples degraded for 6 weeks in a UV chamber. Based on the results, it is possible to recommend the initial treatment of spruce and beech wood with a 2% caffeine solution or 15% solution of FN-NANO® dispersion containing TiO_2 nanoparticles in order to increase the overall service life of transparent acrylic coating in exterior applications.

Author Contributions: The work presented in this paper is a collaborative development by all of the authors. Conceptualisation: M.P. and K.K.; methodology: M.P.; investigation: K.Š., O.D., P.Š., and D.N.; resources:

M.P. and K.K.; data curation: O.S., D.N., and K.Š.; writing—original draft preparation: M.P.; writing—review and editing: K.K.; visualisation, K.Š. All authors have read and agreed to the published version of the manuscript.

Funding: This research was supported by the Czech Science Foundation, Project No. 19-02067S "The effects of methylxanthine-based biocides on the properties of constructional timber" and by Grant "EVA 4.0", No. CZ.02.1.01/0.0/0.0/16_019/0000803 financed by OP RDE.

Conflicts of Interest: The authors declare no conflict of interest.

References

1. Kržišnik, D.; Lesar, B.; Thaler, N.; Humar, M. Influence of natural and artificial weathering on the colour change of different wood and wood-based materials. *Forests* **2018**, *9*, 488. [CrossRef]
2. Feist, W.C. *Weathering and Protection of Wood*; American Wood-Preservers' Association: Kansas City, KS, USA, 1983.
3. Imken, A.A.P.; Brischke, C.; Kögel, S.; Krause, K.C.; Mai, C. Resistance of different wood-based materials against mould fungi: A comparison of methods. *Eur. J. Wood Wood Prod.* **2020**, *78*, 661–671. [CrossRef]
4. Gobakken, L.R.; Lebow, P.K. Modelling mould growth on coated modified and unmodified wood substrates exposed outdoors. *Wood Sci. Technol.* **2010**, *44*, 315–333. [CrossRef]
5. Gaylarde, C.C.; Morton, L.H.G.; Loh, K.; Shirakawa, M.A. Biodeterioration of extrernal architectural paint films—A review. *Int. Biodeterior. Biodegrad.* **2011**, *65*, 1189–1198. [CrossRef]
6. Cogulet, A.; Blanchet, P.; Landry, V. The multifactorial aspect of wood weathering: A review based on a holistic approach of wood degradation protected by clear coating. *BioResources* **2018**, *13*, 2116–2138. [CrossRef]
7. Evans, P.D.; Vollmer, S.; Kim, J.D.W.; Chan, G.; Gibson, S.K. Improving the performance of clear coatings on wood through the aggregation of marginal gains. *Coatings* **2016**, *6*, 66. [CrossRef]
8. Gobakken, L.R.; Westin, M. Surface mould growth on five modified wood substrates coated with three different coating systems when exposed outdoors. *Int. Biodeterior. Biodegrad.* **2008**, *62*, 397–402. [CrossRef]
9. Evans, P.D.; Haase, J.G.; Shakri, A.; Seman, B.M.; Kiguchi, M. The search for durable exterior clear coatings for wood. *Coatings* **2015**, *5*, 830–864. [CrossRef]
10. Cristea, M.V.; Riedl, B.; Blanchet, P. Enhancing the performance of exterior waterborne coatings for wood by inorganic nanosized UV absorbers. *Prog. Org. Coat.* **2010**, *69*, 432–441. [CrossRef]
11. George, B.; Suttie, E.; Merlin, A.; Deglise, X. Photodegradation and photostabilisation of wood—The state of the art. *Polym. Degrad. Stab.* **2005**, *88*, 268–274. [CrossRef]
12. Forsthuber, B.; Schaller, C.; Grüll, G. Evaluation of the photo stabilising efficiency of clear coatings comprising organic UV absorbers and mineral UV screeners on wood surfaces. *Wood Sci. Technol.* **2013**, *2*, 281–297. [CrossRef]
13. Wang, X.; Liu, S.; Chang, H.; Liu, J. Sol-gel deposition of TiO_2 nanocoatings on wood surfaces with enhanced hydrophobicity and photostability. *Wood Fiber Sci.* **2014**, *46*, 109–117.
14. Sandberg, D. Additives in Wood Products—Today and Future Development. In *Environmental Impacts of Traditional and Innovative Forest-Based Bioproducts, Environmental Footprints and Eco-Design of Products and Processes*; Springer Science + Business Media Singapore: Singapore, 2016; pp. 105–172. [CrossRef]
15. Reinprecht, L. *Wood Deterioration, Protection and Maintenance*, 1st ed.; John Wiley & Sons: Hoboken, NJ, USA, 2016; p. 376.
16. Verdier, T.; Coutand, M.; Bertron, A.; Roques, C. Antibacterial Activity of TiO_2 photocatalyst alone or in coatings on *E. coli*: The influence of methodological aspects. *Coatings* **2014**, *4*, 670–686. [CrossRef]
17. Guo, H.; Michen, B.; Burgert, I. Real test-bed studies at the ETH House of Natural Resources–wood surface protection for outdoor applications. *Inf. Constr.* **2017**, *69*, 9. [CrossRef]
18. Salem, M.Z.M.; Mansour, M.M.A.; Mohamed, W.S.; Ali, H.M.; Hatamleh, A.A. Evaluation of the antifungal activity of treated *Acacia saligna* wood with Paraloid B-72/TiO_2 nanocomposites against the growth of *Alternaria tenuissima, Trichoderma harzianum* and *Fusarium culmorum*. *BioResources* **2017**, *12*, 7615–7627. [CrossRef]
19. Singh, T.; Singh, A.P. A review on natural products as wood protectant. *Wood Sci. Technol.* **2012**, *46*, 851–870. [CrossRef]
20. Arora, D.S.; Ohlan, D. In vitro studies on antifungal activity of tea (*Camellia sinensis*) and coffee (*Coffea arabica*) against wood-rotting fungi. *J. Basic Microbiol.* **1997**, *37*, 159–165. [CrossRef]

21. Kwaśniewska-Sip, P.; Cofta, G.; Nowak, P.B. Resistance of fungal growth on Scots pine treated with caffeine. *Int. Biodeterior. Biodegrad.* **2018**, *132*, 178–184. [CrossRef]
22. Kobetičová, K.; Nábělková, J.; Ďurišová, K.; Šimůnková, K.; Černý, R. Antifungal activity of methylxanthines based on their properties. *BioResources* **2020**, *15*, 8110–8120. [CrossRef]
23. Broda, M.; Mazela, B.; Frankowski, M. Durability of wood treated with AATMOS and caffeine—Towards the long-term carbon storage. *Maderas Cienc. Tecnol.* **2018**, *20*, 455–468. [CrossRef]
24. Moya, R.; Rodríguez-Zuniga, A.; Vega-Baudrit, J.; Puente-Urbina, A. Effects of adding TiO_2 nanoparticles to a water-based varnish for wood applied to nine tropical woods of Costa Rica exposed to natural and accelerated weathering. *J. Coat. Technol. Res.* **2016**, *14*, 141–152. [CrossRef]
25. Pánek, M.; Hýsek, Š.; Dvořák, O.; Zeidler, A.; Oberhofnerová, E.; Šimůnková, K.; Šedivka, P. Durability of the exterior transparent coatings on nano-photostabilized English oak wood and possibility of its prediction before artificial accelerated weathering. *Nanomaterials* **2019**, *9*, 1568. [CrossRef] [PubMed]
26. *EN 335:2013 Durability of Wood and Wood-Based Products—Use Classes: Definitions, Application to Solid Wood and Wood-Based Products*; European Committee for Standardization: Brussels, Belgium, 2013.
27. *EN 350:2016 Durability of Wood and Wood-Based Products—Testing and Classification of the Durability to Biological Agents of Wood and Wood-Based Materials*; European Committee for Standardization: Brussels, Belgium, 2016.
28. Pánek, M.; Reinprecht, L. Colour stability and surface defects of naturally aged wood treated with transparent paints for exterior constructions. *Wood Res.* **2014**, *59*, 421–430.
29. Reinprecht, L.; Pánek, M. Effects of wood roughness, light pigments, and water repellent on the color stability of painted spruce subjected to natural and accelerated weathering. *BioResources* **2015**, *10*, 7203–7219. [CrossRef]
30. Gejdoš, M.; Lieskovský, M.; Potkány, M.; Nozdrovický, L. The future perspectives of spruce and fir wood use in selected countries of the Central Europe for wooden constructions. In *Increasing the Use of Wood in the Global Bio-Economy 2018*; Proceeding paper; University of Belgrade: Belgrade, Serbia; WoodEMA: Zagreb, Croatia, 2018; pp. 160–167.
31. Burri, S.; Haeler, E.; Eugster, W.; Haeni, M.; Etzold, S.; Walthert, L.; Braun, S.; Zweifel, R. How did Swiss forest trees respond to the hot summer 2015? *Erde* **2019**, *150*, 214–229. [CrossRef]
32. *EN 927-6:2008 Paints and Varnishes. Coating Materials and Coating Systems for Exterior Wood. Exposure of Wood Coatings to Artificial Weathering Using Fluorescent UV Lamps and Water*; European Committee for Standardization: Brussels, Belgium, 2008.
33. Stearns, E.I. *Colorimetry*, 2nd ed.; Commission Internationale de l'Eclairage: Vienna, Austria, 1986; p. 74.
34. *EN ISO 2813:2015 Paints and Varnishes, Determination of Gloss Value at 20°, 60° and 85°*; European Committee for Standardization: Brussels, Belgium, 2015.
35. ČSN 49 0604:1982 Protection of Wood. In *Methods for Determining the Biocidal Properties of Wood Preservatives*; Office for Standardization and Measurement: Prague, Czech Republic, 1982.
36. Ziglio, A.C.; Sardela, M.R.; Goncalves, D. Wettability, surface free energy and cellulose crystallinity for pine wood (*Pinus* sp.) modified with chili pepper extracts as natural preservatives. *Cellulose* **2018**, *25*, 6151–6160. [CrossRef]
37. Bulian, F.; Graystone, J. *Wood Coatings—Theory and Practice*; Elsevier Science: Amsterdam, The Netherlands, 2009; p. 320. ISBN 978-0444528407.
38. de Meier, M. A Review of Interfacial Aspects in Wood Coatings: Wetting, Surface Energy, Substrate Penetration and Adhesion. COST E18 Final Seminar 2005 16pp. Available online: https://www.researchgate.net/publication/260601859 (accessed on 9 March 2014).
39. Šimůnková, K.; Reinprecht, L.; Nábělková, J.; Kindl, J.; Borůvka, V.; Šobotník, J.; Lišková, T.; Pánek, M. Caffeine—Perspective natural biocide for wood protection in interiors against decaying fungi and termites. **2020**. Prepared for publication.
40. Šimůnková, K.; Zeidler, A.; Schönfelder, O.; Pánek, M. Impact of modification by caffeine on some surface properties of beech wood. In Proceedings of the 9th Hardwood Conference, Sopron, Hungary, 21–22 October 2020. (moved due to pandemic to 2021).
41. Volkmer, T.; Noël, M.; Arnold, M.; Strautmann, J. Analysis of lignin degradation on wood surfaces to create a UV-protecting cellulose rich layer. *Int.. Wood Prod. J.* **2016**, *7*, 156–164. [CrossRef]
42. Müller, U.; Ratzsch, M.; Schwanninger, M.; Steiner, M.; Zobl, H. Yellowing and IR-changes of spruce wood as result of UV-irradiation. *J. Photochem. Photobiol. B Biol.* **2003**, *69*, 97–105. [CrossRef]

43. Sudiyani, Y. Chemical characteristics of surfaces of hardwood and softwood deteriorated by weathering. *J. Wood Sci.* **1999**, *45*, 348–353. [CrossRef]
44. Pánek, M.; Oberhofnerová, E.; Hýsek, Š.; Šedivka, P.; Zeidler, A. Colour stabilization of oak, spruce, larch and Douglas fir heartwood treated with mixtures of nanoparticles dispersions and UV-stabilizers after exposure to UV and VIS-radiation. *Materials* **2018**, *11*, 1653. [CrossRef] [PubMed]
45. Ghosh, M.; Gupta, S.; Kumar, V.S.K. Studies on the loss of gloss of shellac and polyurethane finishes exposed to UV. *Maderas Cienc. Technol.* **2015**, *17*, 39–44. [CrossRef]
46. Pánek, M.; Oberhofnerová, E.; Zeidler, A.; Šedivka, P. Efficacy of hydrophobic coatings in protecting oak wood surfaces during accelerated weathering. *Coatings* **2017**, *7*, 172. [CrossRef]
47. Lie, S.K.; Thiis, T.K.; Vestol, G.I.; Hoibo, O.; Gobakken, L.R. Can existing mould growth models be used to predict mould growth on wooden claddings exposed to transient wetting? *Build. Environ.* **2019**, *152*, 192–203. [CrossRef]
48. Wasserbauer, R. Microbial biodeterioration of electrotechnical insulation materials. *Int. Biodeterior. Biodegrad.* **2004**, *53*, 171–176. [CrossRef]
49. Kwaśniewska-Sip, P.; Bartkowiak, M.; Cofta, G.; Nowak, P.B. Resistance of scots pine (*Pinus sylvestris* L.) after treatment with caffeine and thermal modification against *Aspergillus niger*. *BioResources* **2019**, *14*, 1890–1898. [CrossRef]
50. Fonseca, C.; Ochoa, A.; Ulloa, M.T.; Alvarez, E.; Canales, D.; Zapata, P.A. Poly(lactic acid)/TiO_2 nanocomposites as alternative biocidal and antifungal materials. *Mat. Sci. Eng. C-Biomim.* **2015**, *57*, 314–320. [CrossRef]
51. Požgaj, A.; Chovanec, D.; Kurjatko, S.; Babiak, M. *Štruktúra a Vlastnosti Dreva. (Structure and Properties of Wood)*, 1st ed.; Príroda a.s.: Bratislava, Slovakia, 1993; p. 485.

Publisher's Note: MDPI stays neutral with regard to jurisdictional claims in published maps and institutional affiliations.

coatings

MDPI

Article

Peroxide Post-Treatment of Wood Impregnated with Micronized Basic Copper Carbonate

Rod Stirling [1,*], Gabrielle Boivin [2], Adnan Uzunovic [1], Stacey Kus [1] and John N. R. Ruddick [3]

1 FPInnovations, 2665 East Mall, Vancouver, BC V6T 1Z4, Canada; adnan.uzunovic@fpinnovations.ca (A.U.); stacey.kus@fpinnovations.ca (S.K.)
2 FPInnovations, 1055 rue du P.E.P.S., Québec City, QC G1V 4C7, Canada; gabrielle.boivin@fpinnovations.ca
3 Mychem Wood Protection Consultants Ltd., 1365 Ottawa Avenue, West Vancouver, BC V7T 2H6, Canada; john.n.ruddick@gmail.com
* Correspondence: rod.stirling@fpinnovations.ca; Tel.: +1-604-222-5712

Received: 25 November 2020; Accepted: 10 December 2020; Published: 12 December 2020

Abstract: Wood is vulnerable to significant color changes when used in exterior applications. Some copper-based wood preservatives use colorants to minimize this color change. This paper examines the ability of a peroxide post-treatment to turn wood impregnated with micronized basic copper carbonate ($CuCO_3{\cdot}Cu(OH)_2$) (MBCC) a stable brown color. MBCC-treated wood, with and without peroxide post-treatment, along with associated controls were evaluated for color change, erosion and black-stain fungal resistance after exposure to artificial photo-degradation. The impact of the peroxide treatment on copper leaching was assessed in a laboratory experiment, and the impact on copper reactivity was assessed by electron parametric resonance (EPR) spectroscopy. Peroxide post-treatment of wood pressure impregnated with MBCC was shown to reduce color change by more than 50% compared to controls. Erosion due to photo-degradation and colonization by black-stain fungi were lower in samples treated with MBCC than in untreated controls and were relatively unaffected by peroxide post-treatment. The peroxide post-treatment was associated with increased amounts of mobile copper. This led to increased susceptibility to leaching and to a more than 60% increase in the amount of copper than had reacted with the wood.

Keywords: leaching; micronized basic copper carbonate; peroxide; surface protection; weathering; wood

1. Introduction

The service life of wood products in exterior applications is often limited by loss of appearance rather than loss of structural integrity [1]. As a result, there is a need for wood protection treatments to protect the appearance of the wood as well as against biodegradation by decay and insects. Weathering is a complex series of interactions between moisture, UV and visible light, oxygen, and surface-colonizing fungi [2]. Changes in color arise largely from the photo-degradation of lignin and the colonization by surface-inhabiting black-stain fungi [3].

A wide range of inorganic UV absorbers can be used to improve color stability [4]. This includes copper, which is already present in many residential preservative formulations. Copper-based wood preservatives are known to photo-stabilize lignin and slow the rate of surface degradation and color change from weathering [5–8]. One of the most widely used residential wood protection treatments in North America is micronized copper azole (MCA), which includes micronized basic copper carbonate ($CuCO_3{\cdot}Cu(OH)_2$) (MBCC) as the primary biocide. Treatment with MBCC gives wood a pale blue/green color. Commercial products often add a colorant to give the MBCC-treated wood a more natural-looking brown color [9]. For exterior preserved wood, iron oxides dominate the market due to their low cost, the wide range of colors that are possible, their photo-protective effects, their compatibility with preservative formulations and their positive health and environmental profiles [10]. However, the color still fades over time.

It was observed that, in service, wood treated with copper-based preservatives tends to go from green/blue to a brown color as the surface of the wood oxidizes [11]. The use of peroxide to generate stable color complexes in wood impregnated with transitional metal compounds has been reported by Auger [12]. In the present work, we explored this approach using wood impregnated with MBCC followed by a peroxide post-treatment. The aim was to understand the efficacy of this approach in yielding a brown, photo-stable wood surface with the potential to eliminate or reduce the need for the addition of colorants.

MBCC-treated wood with a peroxide post-treatment could potentially also affect preservative efficacy. MBCC has proven to be an effective wood preservative due to the slow solubilization of MBCC and reaction of this solubilized copper with the wood cell wall [13–15]. Oxidation of the wood surface was hypothesized to increase the amount of reacted copper, which could potentially increase biological efficacy.

2. Materials and Methods

These experiments assess the performance of wood pressure-treated with MBCC, with and without a peroxide post-treatment. Specimens were evaluated for their resistance to color change and erosion following accelerated photo-degradation, and their resistance to disfigurement by artificially inoculated and incubated black-stain fungi under optimal conditions. An additional experiment examined the impact of the peroxide post-treatment on copper leaching and on the formation of reacted copper in the wood.

Red pine (*Pinus resinosa*) sapwood was used as a wood substrate throughout these experiments. MBCC was obtained from Timber Specialties Co. (Campbellville, ON, Canada). An iron oxide-based colorant formulation was used as a reference in the accelerated photo-degradation experiment. The MBCC and the iron oxide-based colorant were applied to wood by vacuum-pressure impregnation to target gauge concentrations of 4.0 kg of MBCC per cubic meter of wood and 2.0 kg of iron oxide-based colorant per cubic meter of wood, respectively.

The peroxide post-treatment consisted of a one-minute dip in a 20% solution of aqueous hydrogen peroxide adjusted to pH 6 with a dilute solution of sodium hydroxide at room temperature on air-dried samples. The 20% solution was prepared from a 30% hydrogen peroxide concentrate (Fisher Scientific, Ottawa, ON, Canada, ACS Reagent Grade). The conditions described above were determined based on a series of tests to optimize the peroxide treatment conditions. These tests are described in Appendix A. Factors evaluated included peroxide concentration, pH, storage time, drying and the impact of wood type (heartwood vs. sapwood).

2.1. Accelerated Photo-Degradation

Assessment of resistance to accelerated photo-degradation included four treatments groups each with six replicates. These included the MBCC treatment, with and without peroxide post-treatment, the iron oxide-based colorant as a reference product, and untreated red pine sapwood as a control. Specimen size was 75 mm long × 75 mm tangential × 12 mm radial wood fiber direction.

The sample color was evaluated with a SP60 spectrophotometer (X-Rite, Grand Rapids, MI, USA) using CIE L^*a^*b color space. L^* represents lightness on a scale from 0 to 100. Negative a^* values indicate degree of greenness, while positive values indicate degree of redness. Negative b^* values indicate degree of blueness, while positive values indicate degree of yellowness. CIE $L^*a^*b^*$ data were used to calculate ΔL^* (change in lightness), Δa^* (change in green-red color), Δb^* (change in blue-yellow color), and ΔE^* which represents the total color change according to Equation (1).

$$\Delta E^* = \sqrt{(L_2^* - L_1^*)^2 + (a_2^* - a_1^*)^2 + (b_2^* - b_1^*)^2} \quad (1)$$

A small amount of vinyl polysiloxane (Aquasil Ultra Smart Wetting®, Dentsply Caulk, Montreal, QC, Canada), approximately 10 mm × 20 mm, was stuck to three samples from each treatment group. This was used to provide an unexposed reference patch from which to measure erosion following Q Sun exposure. After Q-Sun exposure, the silicone was removed from the surface. A 3D profilometer, Contour GT-K1, (Veeco Instruments, Plainview, NY, USA) was used to measure the erosion of the exposed area compared to the unexposed sections. A total of six measurements were taken per sample (three samples/series of treatment) for a total of eighteen replicates per series of treatment.

Samples were placed in a Q-Sun chamber (Q-Lab Corporation, Westlake, OH, USA) for accelerated photo-degradation. The samples were exposed using the ASTM G-155 cycle 1 method [16]. This cycle includes 102 min of exposure to light at 0.35 W/m^2 irradiance at 340 nm and 63 °C for black panels. The chamber air was maintained at 43 °C and 30% relative humidity. This was followed by 18 min of exposure to water and light at 0.35 W/m^2 irradiance at 340 nm and 63 °C for black panel. Color measurements were repeated after 1000 h of exposure.

2.2. Black-Stain Fungal Resistance

After inspection at 1000 h of exposure, samples from the accelerated photo-degradation experiment were exposed to an additional 500 h of exposure in the Q-Sun. These samples (exposed for a total of 1500 h in the Q-Sun chamber) were cut into four 35 mm × 35 mm × 12 mm subsamples. In addition, red pine sapwood that had not been exposed in the Q-Sun was included as control, and red pine sapwood pressure impregnated with a solution of 1% propiconazole, and 0.5% 3-iodo-2-propynyl butyl carbamate (IPBC) was included as an additional reference known to control black-stain fungi [17]. Twenty-four replicates (6 replicates of each treatment × 4 subsamples) were produced for each treatment. Group A included two isolates with typical *Aureobasidium pullulans* morphology that were isolated from coated wood in Vancouver, BC, Canada. Group B included two isolates with darker and more yeast-like colony morphology that were slower growing on nutrient plates than group A and were isolated from coated wood in eastern Canada. Group C, isolated from both western and eastern Canada, included a mix of ten isolates of various colony types collected from coated wood affected by black stain after only 2 years of in service exposure. Group D was sprayed with distilled water and used as an uninoculated reference. All subsamples were edge sealed with a marine epoxy (Intergard 740, International Marine Coatings, Singapore) and sterilized by ion beam irradiation with two passes at 17 kGy (Iotron Industries, Port Coquitlam, BC, Canada). Prior to inoculation, the samples were aseptically placed in test chambers with 2 mL of sterile distilled water and left to incubate at 22.5 °C for 48 h to ensure a moist surface for inoculation. Test chambers are described in [17].

To prepare a spore suspension for inoculum groups A and B, a solution of distilled water with Tween® 20 (Sigma-Aldrich, Oakville, ON, Canada), a polysorbate-based non-ionic surfactant, was used to gently remove spores from the surface of two actively growing plates by gently shaving the sporulating mycelium using a sterile scalple of each chosen isolate and making up to a final volume of 200 mL. For inoculum C, the solution of sterile water with Tween® 20 was again used to remove spores from the surface of one actively growing plate of each of the ten chosen isolates and the final volume made to 400 mL. The solutions were blended with three short pulses in a Waring commercial blender. The inoculum was then filtered through a prewashed sterile glass wool-lined glass funnel to remove large particles that could plug the air brush. Haemocytometer average counts of spore suspensions for inoculum A, B and C were 1.1 × 10^6, 0.54 × 10^6 and 2.2 × 10^6 spores per millilitre of the 200 and 400 mL spore suspensions, respectively. The inoculum was applied using an IWATA Eclipse (Anest Iwata-Medea, Inc., Portland, OR, USA) HP-BCS airbrush, 5 mL of inoculum was used to evenly coat twelve test pieces at a time. Inoculum sprayed on control plates (1% malt extract agar) developed a healthy growth of black-stain fungi for each inoculum group after a few days of incubation, while plates sprayed with just water inoculum had no growth. Test samples were incubated at 22.5 °C in the dark for six weeks. A weekly in situ spritzing with sterile distilled water was applied by a gentle spray over the sample surface to maintain a suitable moisture content.

Samples were inspected after two, four and six weeks of incubation based on the visual inspection scale described in [18]. However, it was noticed that some fungi altered the color of the surface but had no visible growing mycelia or sporing structures on the surface. For such samples, the discoloration observed on the surface was compared to uninoculated control samples, and a rating was given based on the amount of discoloration present and not only on visible fungal growth on the surface. Ratings of 0 indicated no stain, while ratings of 5 indicated intense and widespread staining of the sample.

2.3. Copper Leaching

Further studies were conducted to assess the impact of the peroxide treatment on the leachability of copper from the treated wood. Thirty samples of red pine (*Pinus resinosa*) sapwood were cut to the dimensions of 50 mm long × 15 mm radial × 25 mm tangential wood fiber direction and were conditioned to constant mass at 20 °C and 65% relative humidity (RH). Twenty samples were pressure treated with an aqueous suspension of MBCC to an average gauge retention of 3.2 kg of MBCC per cubic meter of wood. Specimens were reconditioned to constant weight at 20 °C and 65% RH following treatment. Ten specimens were dip treated in a 20% solution of hydrogen peroxide (pH 6) and ten were dip treated in water for five minutes. The wood specimens were then reconditioned to constant weight and evaluated for copper leaching using the immersion cycle specified in section 7.3 of the OECD Guidance document [19]. The mass of the samples was recorded, and the samples were completely immersed in water for 120 min, removed from the water and allowed to drain for 10 s, allowing run-off to return to the water. The leachate samples were collected and stored in the freezer. The test specimens were weighed again and then left to dry at laboratory temperature until the next immersion day. Immersion occurred on days: 1, 3, 7, 9, 11, 14, 16 and 22. Leachate samples were then sent to Maxxam Analytics (Burnaby, BC, Canada) for copper analysis by Inductively Coupled Plasma Atomic Emission Spectroscopy (ICP-AES). After all immersion cycles had been completed the samples were left to dry at laboratory temperature. Once dry the leached samples were milled to pass through a 40-mesh screen and pressed into pellets. The pellets were measured for total copper using an ARL™ QUANT'X EDXRF Spectrometer (ThermoFisher Scientific, Waltham, MA, USA) [20].

2.4. Copper Characterization

To assess the impact of the peroxide treatment on the form of copper present in the treated wood, six red pine sapwood specimens (25 mm × 50 mm × 15 mm) were pressure impregnated with MBCC. Three of these specimens were then dip treated in a 20% aqueous solution of hydrogen peroxide. All specimens were then milled to pass through a 40-mesh screen. The resulting sawdust was analyzed by x-ray fluorescence spectroscopy for total copper. The remaining sawdust from each specimen was analyzed by Electron Paramagnetic Resonance (EPR) spectroscopy at the University of British Columbia (Vancouver, BC, Canada) using the methods developed by Xue et al. [21]. A Bruker Elexsis E500 series continuous wave EPR (Bruker BioSpin GmbH, Rheinstetten, Germany) was used for the analysis. The operating parameters were 77 K, at a frequency 940 GHz (X-band), 100 KHz field modulation and 1 G modulation amplitude. Frequency calibration was independently verified using 2,2-diphenyl-1-picrylhydrazyl. The spectral simulation was achieved using SIMFONIA (Bruker Biospin GmbH, Rheinstetten, Germany). Integration was OriginPro8 (Origin Lab). Baseline correction was applied before calculating the area under the curve. The sample tube was packed with wood sawdust to a height greater than 100 mm so that it exceeded the cavity height, thereby minimizing variation caused by the sample content.

3. Results

3.1. Accelerated Photo-Degradation

Average color change after 1000 h of Q-Sun exposure was calculated for each treatment group (Figure 1). The MBCC-treated wood became darker and less green. The peroxide post-treatment greatly reduced these color changes, as the wood was already darker and less green prior to Q-Sun exposure. The iron oxide-based reference treatment became less red and less yellow after Q-Sun exposure. The untreated control became darker and much less yellow. The total color change (ΔE^*) was similar between the peroxide post-treatment and the iron oxide-based colorant reference, and less than the MBCC without the peroxide post-treatment and the untreated control.

Erosion measurements were quite variable but did show a significant reduction in erosion for all samples treated with copper compounds and the proprietary iron oxide colorant (Figure 2). This is consistent with the ability of copper to photo-stabilize wood [5–7,22]. Erosion was similar in MBCC-treated samples with and without peroxide post-treatment.

Figure 3 shows the samples before and after 1000 h of Q-Sun exposure. The discolored spot on three of the samples from each group is from the blue tack that was used to create a reference spot for erosion measurement. MBCC-treated wood changed from a pale green to a medium brown color. This is consistent with field observations of such material [23]. The peroxide treatment resulted in an initial brown-green color that changed to the same medium brown color as the MBCC treatment.

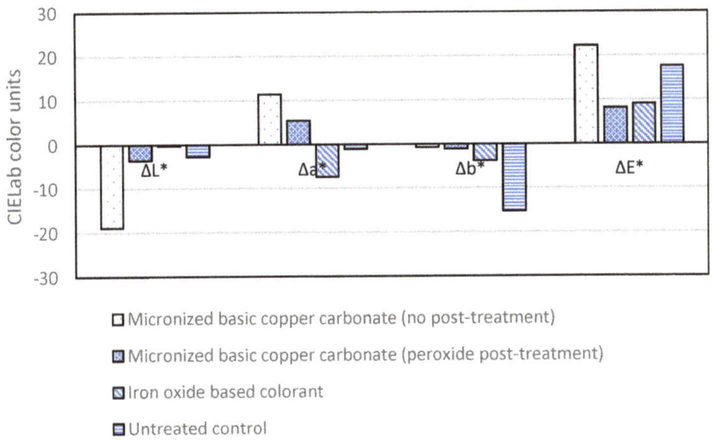

Figure 1. Color change after 1000 h of Q-Sun exposure.

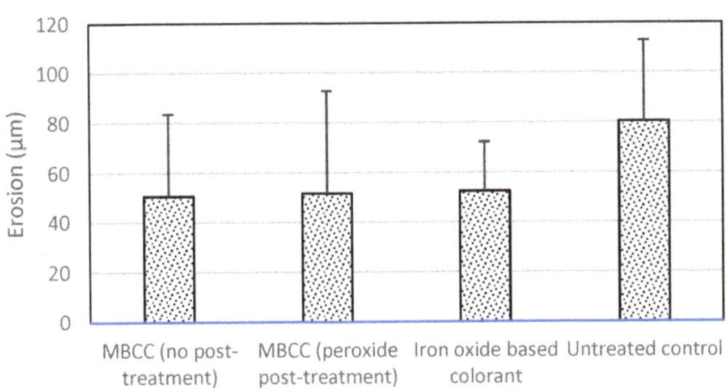

Figure 2. Average and standard deviation of surface erosion after 1000 h of Q-Sun exposure of samples treated with micronized basic copper carbonate (MBCC) with and without peroxide post-treatment, an iron oxide-based colorant, and an untreated control.

Figure 3. Red pine sapwood before and after 1000 h of accelerated photo-degradation: (**a**) treated with micronized basic copper carbonate, (**b**) treated with micronized basic copper carbonate after 1000 h of accelerated photo-degradation, (**c**) treated with micronized basic copper carbonate with a peroxide post-treatment, (**d**) treated with micronized basic copper carbonate with a peroxide post-treatment after 1000 h of accelerated photo-degradation, (**e**) treated with an iron oxide-based colorant, (**f**) treated with an iron oxide-based colorant after 1000 h of accelerated photo-degradation, (**g**) untreated, and (**h**) untreated after 1000 h of accelerated photo-degradation.

3.2. Black-Stain Fungal Resistance

Black-stain fungi aggressively attacked most samples and quickly grew, causing darkening or masking of the original wood color (Figure 4). Note that the white material on some surfaces is epoxy resin from the edge seal. No color changes were observed on the uninoculated controls (Group D). A rating of 5 was assigned to the majority of samples of unweathered and weathered inoculated pine controls after only two weeks of incubation (Figure 5). The unweathered IPBC/propiconazole reference performed well and prevented any fungal growth with an average rating of 0 throughout the test for all inoculum types. The MBCC treatment was associated with lower black-stain ratings than untreated controls. Average stain ratings were near zero for inoculum group A. However, growth of the strains in inoculum groups B and C was evident on the MBCC-treated wood, with or without peroxide post-treatment. The peroxide post-treatment was associated with lower black-stain ratings than MBCC with no post-treatment for inoculum group B, but was similar for group C. MBCC is used to protect wood from fungal decay. This does not include staining fungi, though these data suggest a possible beneficial effect.

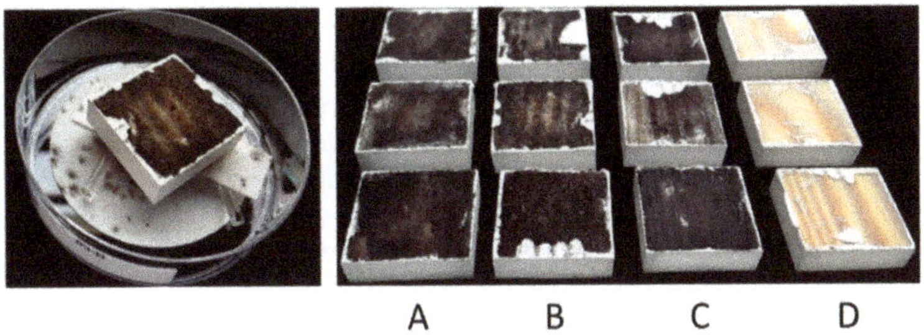

Figure 4. Test assembly showing unweathered red pine sapwood covered with black-stain fungi after two weeks of incubation (**left**); unweathered red pine sapwood samples after six weeks of incubation with inoculum groups A, B, C and D (uninoculated) (**right**).

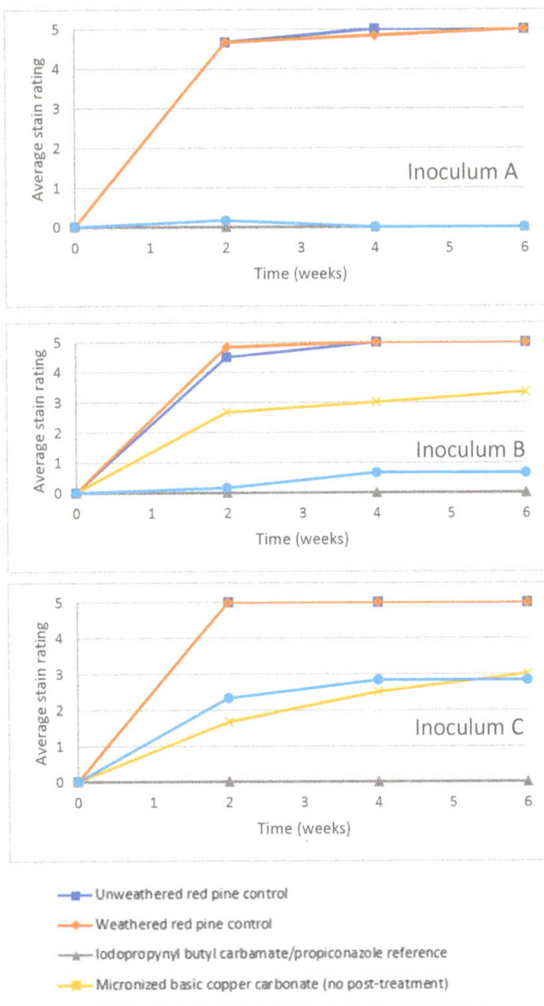

Figure 5. Average stain ratings over a six-week incubation for specimens exposed to strains of *Aureobasidium pullulans* from inoculum groups (**A–C**).

3.3. Copper Leaching

Copper leaching from the wood treated with MBCC was very low and similar to the untreated control (Figure 6). This is consistent with previous work demonstrating the low leachability of copper from wood impregnated with MBCC [24]. In contrast, copper leaching was much greater in the samples post-treated with peroxide, though this tapered off substantially after 22 days. During the leaching experiment, it was observed that water uptake during immersion was much greater in the peroxide post-treatment group (average 6.4 g) than in the no post-treatment group (average 3.4 g). This increase in water uptake may explain the greater degree of leaching observed.

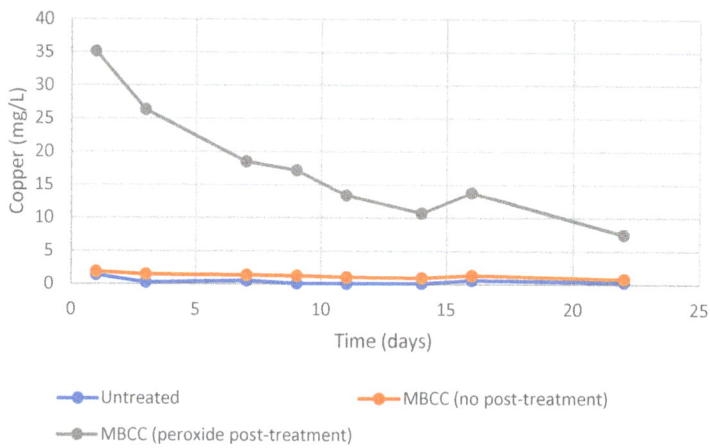

Figure 6. Concentration of copper in leachates from wood impregnated with micronized basic copper carbonate, with and without peroxide post-treatment.

3.4. Copper Characterization

The total copper content in the MBCC-treated wood measured by X-ray fluorescence is shown in Table 1. Total copper concentration averaged 5.6 mg/g in material with no post-treatment and 5.0 mg/g in material with the peroxide post-treatment. Reacted copper concentration averaged 2.62 mg/g in material with no post-treatment and 4.24 mg/g in material with the peroxide post-treatment. These data demonstrate an association between the peroxide post-treatment and increased levels of reacted copper. The reacted copper content in the wood treated with micronized copper carbonate is typical of values previously reported [21,25]. The peroxide treatment may generate more reactive sites in the wood, leading to enhanced reaction between the wood and the micronized basic copper carbonate. The peroxide reaction with lignin is known to oxidize ketonic carbonyl groups [26].

The EPR spectral parameters provide information on the electronic structure of paramagnetic copper complexes in wood. There was very little variation between material with and without peroxide post-treatment, indicating that the compounds being created during the post-treatment have very similar configuration and bonding as the reacted copper (Table 2). The A_z parameter is consistent with a copper–oxygen bonding typical of the copper carboxylate bonding formed in wood [21].

Table 1. Average total copper measured by X-ray fluorescence and reacted copper measured by EPR in MBCC-treated wood, with and without a peroxide post-treatment.

Treatment	Total Copper (mg/g)	Reacted Copper (mg/g)
No post-treatment	5.57	2.62
Peroxide post-treatment	5.04	4.24

Table 2. Average EPR spectral data from MBCC-treated wood, with and without a peroxide post-treatment.

Treatment	g_x	g_y	g_z	A_x	A_y	A_z
No post-treatment	2.078	2.084	2.373	5	45	131
Peroxide post-treatment	2.076	2.084	2.375	5	46	133

4. Discussion

The color change and erosion data confirm the value of copper in wood protection for its ability to partially photo-stabilize wood as has been previously been reported [5–8]. The improved resistance to black-stain fungal colonization of the MBCC-treated samples is consistent with previous work showing that copper can inhibit these organisms. Plasma-deposited copper has been found to inhibit the growth of *A. pullulans* [27]. However, some copper tolerance has also been observed with these fungi [28]. This may indicate why only some growth was observed. The copper tolerance of the fungal strains used has not been otherwise assessed. The natural looking mid-brown color produced by the peroxide post-treatment of MBCC impregnated wood suggests that an attractive and fade resistant brown color at point of sale could be achieved. This could potentially reduce or eliminate the need for the addition of colorants, which are widely used in wood treated with MBCC-containing preservatives [9,10].

The performance against different black-stain inocula varied considerably, though the performance trends were similar. This highlights the importance of carefully choosing test isolates and in evaluating field-collected, fresh isolates, not mutilated through long storage and inclusion of more than one strain to better understand performance. The optimized growth conditions of this accelerated set up led to very rapid attack that would not be observed in field studies. Field studies on the growth of black-stain fungi on MBCC-treated wood would be useful to better understand the effect of the MBCC treatment. Co-biocides, such as the propiconazole used in micronized copper azole type C, would likely also contribute to efficacy in the field.

This study did not examine the use of peroxide dip treatment before pressure treatment with MBCC. Given the increased water uptakes observed in the leaching experiment, peroxide pre-treatment could potentially improve preservative solution uptake. This could make it easier for treaters to meet preservative penetration and retention requirements and lead to better treated products and/or more rapid treatment schedules. A more detailed assessment of the interaction between peroxide and copper carbonate should be conducted to optimize the treatment so that it maximizes reacted copper and minimizes leaching while still producing the desired brown color.

Peroxide is a common ingredient in deck cleaning products. These data suggest that treatments may result in increased copper mobility and increased levels of reacted copper. Bleach-based deck cleaners have been associated with increased copper leaching from CCA-treated decking [29]. The authors of that study note that amount of copper leached would not be a concern in a residential context. It is unclear to what extent a peroxide-based deck cleaner would mobilize copper from wood treated with a MBCC-containing preservative.

The potential impact of increased copper leaching and reactivity on long-term durability should be investigated. The increased water uptake and copper leaching observed in this study suggest a potential loss of durability. However, the increased amount of reacted copper suggests a potential improvement in durability. Previous work has found that copper from MBCC-treated wood is resistant to leaching and that the leached wood retained its resistance to decay fungi [30].

EPR has been used previously to study the reaction chemistry of copper-based wood preservatives in wood [15,31]. While several analytical tools are useful for quantifying copper in organic matrices, most techniques cannot provide information on the copper species present or the bonding of the copper to the wood components. One technique that can identify copper–ligand bonding is EPR. However, it too is limited to copper species that are paramagnetic. While this is usually not a problem it could be limiting, in that it cannot detect species for example that are antiferromagnetic or Cu(I) species. Studies by Piesach and Blumberg [32] have shown that the unpaired electron chemical shift in mononuclear

$S = 1/2$ Cu(II) species and hyperfine coupling to the $I = 3/2$ copper nucleus along axial direction (the g_z and A_z respectively) are most sensitive to electronic and geometric perturbations in the bonding to the copper ion. All the spectra obtained in this study were very similar. They all exhibited strong g_z values of ≈2.373, and A_z tensors of ≈131 G which are typical of copper-wood complexes formed with only copper-oxygen bonding. This supported the hypothesis that all of the copper complexes being formed both with, and without peroxide treatment, were identical in the coordination chemistry of the copper. This would suggest that the peroxide treatment is forming carboxylic acid functional groups which are then able to mobilize copper from the remaining basic copper carbonate. The high solubility of the copper compounds formed suggests that, unlike the natural carboxylic acid functional groups present in wood, those formed by the peroxide post-treatment may be wood-degradation products. In the basic copper carbonate present in the MBCC, the copper is antiferromagnetically coupled in the solid state, and so is EPR silent.

5. Conclusions

Peroxide post-treatment of wood pressure impregnated with MBCC may help to minimize color change due to photo-degradation. Erosion due to photo-degradation and colonization by black-stain fungi were similar in samples impregnated with MBCC, regardless of the peroxide post-treatment. The peroxide post-treatment was associated with increased copper leaching, but also with increased reacted copper.

Author Contributions: Conceptualization and design, R.S., G.B., and A.U.; photostability testing, G.B.; black-stain testing, A.U. and S.K.; EPR analysis, J.N.R.R.; draft manuscript preparation, R.S. and G.B.; review and editing, R.S., G.B., A.U., S.K., and J.N.R.R. All authors have read and agreed to the published version of the manuscript.

Funding: This project was financial supported by Natural Resources Canada under a contribution agreement with FPInnovations.

Acknowledgments: The authors thank Timber Specialties Co. for providing the micronized basic copper carbonate used in this experiment and Pierre Kennepohl (University of Calgary, formerly University of British Columbia) for providing EPR access. The technical support and guidance from Mathieu Gosselin, Ashley Hook, Paul Morris, Stéphane Thibeault and Daniel Wong is gratefully acknowledged.

Conflicts of Interest: The authors declare no conflict of interest. The funders had no role in the design of the study; in the collection, analyses, or interpretation of data; in the writing of the manuscript, or in the decision to publish the results.

Appendix A. Peroxide Treatment Optimization

White spruce (*Picea glauca*) was cut into $1 \times 10 \times 50$ mm^3 strips from sapwood and dip treated with a 1% solution of MBCC intended to simulate the surface of lumber pressure impregnated with a micronized copper containing preservative. These treated strips were used to examine the color change associated with oxidizing the copper when exposed to hydrogen peroxide. Initially a solution of 30% hydrogen peroxide was applied to dried copper treated spruce test strips, and a very dark brown color was produced, confirming that oxidation was possible. To produce the desired brown hue, testing was conducted to optimize hydrogen peroxide concentration and pH. Higher pH was obtained by addition of dilute sodium hydroxide. Lower pH was obtained by addition of dilute hydrochloric acid. Reaction time, drying method as well as influences from species and substrates were also explored.

The initial screening test looked at the effect of hydrogen peroxide concentration on color (Figure A1). A 10% solution (pH 6) resulted in an uneven light brown color. Treatments with 20% (pH 5) and 30% (pH 4) solutions resulted in a more uniform medium brown color.

Figure A1. The effect of hydrogen peroxide concentration on the color of spruce treated with MBCC (left to right: untreated, MBCC treated (no peroxide), MBCC-treated 10% peroxide at pH 6, MBCC treated 20% peroxide at pH 5, 30% peroxide at pH 4).

When treated with a 20% solution of hydrogen peroxide, pH 6 was found to best yield a medium brown color (Figure A2). Treatment with a 10% solution at varying pH levels resulted in only light brown coloration at all pH levels and an uneven tone across the samples.

Figure A2. The effect of peroxide solution pH of the color of spruce treated with MBCC (top 20% peroxide, left to right pH 4, pH 6 and pH 8; bottom 10% peroxide, left to right pH 4, pH 6, pH 8).

The effect of storage time was assessed on MBCC-treated samples exposed to 10% hydrogen peroxide (pH 6) (Figure A3). Samples were held wet at different time frames then air dried, left wet or oven dried at 100°C before the hydrogen peroxide was applied. For hold time, a moderate browning was observed in all samples, suggesting that there was no improved color development with a longer storage time.

Figure A3. The effect of wet storage time on the color of spruce treated with MBCC and 10% peroxide (left to right 0 h, 1 h and 22 h hold before oven drying).

There was an impact in color development when different drying methods were employed. Little difference was observed between air drying and oven drying the samples before treating them with hydrogen peroxide, as similar colors were produced with the secondary treatment. However, it was noticed that the initial copper treatment washed off when samples were treated with peroxide while still wet. This was evidenced by the peroxide solution turning brown and the treated samples having a slightly muted brown color (Figure A4).

Figure A4. The effect of oven drying on the color of spruce treated with MBCC and 15% peroxide.

Samples of mixed sapwood and heartwood were treated with 20% peroxide at pH 6 and 8 (Figure A5). The heartwood/sapwood boundary was visually evident in the basic copper carbonate treated samples and remained visible after exposure to peroxide. However, both wood types turned a similar mid-brown color.

Based on these experiments, the optimum conditions were determined to be 20% hydrogen peroxide at pH 6 on dried copper treated samples, producing a consistent rich brown color across the samples. Wet storage time and substrate did not have major effects, while not drying samples before hydrogen peroxide treatments removed some of the copper treatment applied affecting the color produced.

Figure A5. The effect pine sapwood and heartwood on the color of wood treated with MBCC and exposed to 20% hydrogen peroxide at pH 8 (top) and pH 6 (bottom).

References

1. Sandak, A.; Sandak, J.; Dimitriou, A.; Burud, I.; Thiis, T.K.; Gobakken, L.R.; Ormondroyd, G.A.; Kraniotis, D. Assessment and monitoring of aesthetic appearance of building biomaterials during the service life. *WIT Trans. Ecol. Environ.* **2017**, *226*, 527–536. [CrossRef]
2. Evans, P.; Chowdhury, M.J.; Mathews, B.; Schmalzl, K.; Ayer, S.; Kiguchi, M.; Kataoka, Y. Weathering and surface protection of wood. In *Handbook of Environmental Degradation of Materials*; Kutz, M., Ed.; William Andrew Publishing: Norwich, NY, USA, 2005; pp. 277–297. [CrossRef]
3. Feist, W.C. Outdoor wood weathering and protection. In *Archaeological Wood: Properties, Chemistry, and Preservation*; Advances in Chemistry; Rowell, R.M., Barbour, R.J., Eds.; American Chemical Society: Washington, DC, USA, 1990; Volume 225, pp. 263–298. [CrossRef]
4. Blanchard, V.; Blanchet, P. Color stability for wood products during use: Effects of inorganic nanoparticles. *BioResources* **2011**, *6*, 1219–1229.
5. Jin, L.; Archer, K.; Preston, A. Surface Characteristics of Wood Treated with Various AAC, ACQ and CCA Formulations After Weathering. International Research Group on Wood Preservation. Document No. IRG/WP/2369. IRG Secretariat, Stockholm Sweden. 1991. Available online: https://www.irg-wp.com/search-irg-docs.html (accessed on 20 November 2020).
6. Liu, R.; Ruddick, J.N.R.; Jin, L. The Influence of Copper (II) Chemicals on the Weathering of Treated Wood, 1: ACQ Treatment of Wood on Weathering. International Research Group on Wood Preservation. Document No. IRG/WP/94-30040. IRG Secretariat, Stockholm, Sweden. 1994. Available online: https://www.irg-wp.com/search-irg-docs.html (accessed on 20 November 2020).
7. Zhang, J.; Kamdem, D.P.; Temiz, A. Weathering of copper-amine treated wood. *Appl. Surf. Sci.* **2009**, *256*, 842–846. [CrossRef]
8. Ozgenc, O.; Hiziroglu, S.; Yildiz, U.C. Weathering properties of wood species treated with different coating applications. *BioResources* **2012**, *7*, 4875–4888. [CrossRef]

9. Zhang, J.; Ziobro, R. Micronized Pigment Technology for Pressure Treatment. Proceedings Canadian Wood Preservation Association, 34, 2013. Available online: https://www.cwpa.ca/publications/ (accessed on 20 November 2020).
10. Archer, K.; Sturdivant, R.; Schauwecker, C. Pressure Treatment of Lumber with Colorant Additives. Proceedings Canadian Wood Preservation Association, 34, 2013. Available online: https://www.cwpa.ca/publications/ (accessed on 20 November 2020).
11. Grelier, S.; Castellan, A.; Kamdem, D.P. Photoprotection of copper-amine-treated pine. *Wood Fiber Sci.* **2000**, *32*, 196–202.
12. Auger, S.B. Mineral Stains for Wood and Other Substrates. US Patent 2007/0011820 A1, 12 September 2007.
13. Stirling, R.; Drummond, J.; Zhang, J.; Ziobro, R.J. Micro-Distribution of Micronized Copper in Southern Pine. International Research Group on Wood Protection. Document No. IRG/WP/08-30479. IRG Secretariat, Stockholm, Sweden. 2008. Available online: https://www.irg-wp.com/search-irg-docs.html (accessed on 20 November 2020).
14. Matsunaga, H.; Kiguchi, M.; Evans, P.D. Microdistribution of copper-carbonate and iron oxide nanoparticles in treated wood. *J. Nanoparticle Res.* **2009**, *11*, 1087–1098. [CrossRef]
15. Xue, W.; Kennepohl, P.; Ruddick, J.N.R. Investigation of copper solubilization and reaction in micronized copper treated wood by electron paramagnetic resonance (EPR) spectroscopy. *Holzforschung* **2012**, *66*, 889–895. [CrossRef]
16. *ASTM G-155 Standard Practice for Operating Xenon Arc Light Apparatus for Exposure of Non-Metallic Materials*; ASTM International: West Conshohocken, PA, USA, 2013.
17. Stirling, R.; Uzunovic, A.; Morris, P.I. Control of black stain fungi with biocides in semitransparent wood coatings. *For. Prod. J.* **2011**, *61*, 359–364. [CrossRef]
18. *AWPA E24-16 Laboratory Method for Evaluating the Mold Resistance of Wood-Based Materials: Mold Chamber Test*; American Wood Protection Association: Birmingham, AL, USA, 2016.
19. OECD Guidance on the Estimation of Emissions from Wood Preservative—Treated Wood to the Environment: For Wood held in Storage After Treatment and for Wooden Commodities that are not covered and are not in Contact with Ground. Organization for Economic Co-operation and Development, Paris, France. 2008. Available online: https://echa.europa.eu/documents/10162/16908203/pt8_oecd_guidance_emission_estimation_treated_Wood_en.pdf/97bc4a98-abf6-4a3a-a448-88681a17dd52 (accessed on 23 July 2009).
20. AWPA A9-16. *Standard Method for Analysis of Treated Wood and Treating Solutions by X-ray Spectroscopy*; American Wood Protection Association: Birmingham, AL, USA, 2016.
21. Xue, W.; Kennepohl, P.; Ruddick, J.N.R. Quantification of mobilized copper (II) levels in micronized copper-treated wood by electron paramagnetic resonance (EPR) spectroscopy. *Holzforschung* **2013**, *67*, 815–823. [CrossRef]
22. Deka, M.; Humar, M.; Rep, G.; Kričej, B.; Šentjurc, M.; Petrič, M. Effects of UV light irradiation on colour stability of thermally modified, copper ethanolamine treated and non-modified wood: EPR and DRIFT spectroscopic studies. *Wood Sci. Technol.* **2008**, *42*, 5–20. [CrossRef]
23. Stirling, R.; Wong, D. The Impact of Coatings on the Service Life of Wood Decking. International Research Group on Wood Protection. Document No. IRG/WP/18-20635. IRG Secretariat, Stockholm, Sweden. 2008. Available online: https://www.irg-wp.com/search-irg-docs.html (accessed on 20 November 2020).
24. Zhang, J.; Ziobro, R.J. Micronized Copper Preservative Systems: Observations on the Release of Cupric ion (Cu2+) from Treated Wood and Performance against Wood Decay Fungi. International Research Group on Wood Protection. Document No. IRG/WP/09-30519. IRG Secretariat, Stockholm, Sweden. 2009. Available online: https://www.irg-wp.com/search-irg-docs.html (accessed on 20 November 2020).
25. Xue, W.; Kennepohl, P.; Ruddick, J.N.R. Reacted copper (II) concentrations in earlywood and latewood of micronized copper-treated Canadian softwood species. *Holzforschung* **2015**, *69*, 509–512. [CrossRef]
26. Kadla, J.F.; Chang, H.M. The reactions of peroxides with lignin and lignin model compounds. In *Oxidative Delignification Chemistry, ACS Symposium Series 785*; American Chemical Society: Washington DC, USA, 2001; pp. 108–129. [CrossRef]
27. Gascón-Garrido, P.; Thévenon, M.F.; Mainusch, N.; Militz, H.; Viöl, W.; Mai, C. Siloxane-treated and copper-plasma-coated wood: Resistance to the blue stain fungus *Aureobasidium pullulans* and the termite *Reticulitermes flavipes*. *Int. Biodeterior. Biodegrad.* **2017**, *120*, 84–90. [CrossRef]

28. Gadd, G.M. Effect of copper on *Aureobasidium pullulans* in solid medium: Adaptation not necessary for tolerant behaviour. *Trans. Br. Mycol. Soc.* **1984**, *82*, 546–549. [CrossRef]
29. Gress, J.; De Oliveira, L.M.; Da Silva, E.B.; Lessl, J.M.; Wilson, P.C.; Townsend, T.; Ma, L.Q. Cleaning-induced arsenic mobilization and chromium oxidation from CCA-wood deck: Potential risk to children. *Environ. Int.* **2015**, *82*, 35–40. [CrossRef] [PubMed]
30. Xu, G.; Cappellazzi, J.; Konkler, M.J.; Morrell, J.J. Effect of Multiple Leaching Cycles on Decay Resistance of Micronized Copper Azole–Treated Southern Pine Sapwood. *For. Prod. J.* **2020**, *70*, 178–181. [CrossRef]
31. Xue, W.; Ruddick, J.N.R.; Kennepohl, P. Solubilisation and chemical fixation of copper(II) in micronized copper treated wood. *Dalton Trans.* **2016**, *45*, 3679–3686. [CrossRef] [PubMed]
32. Peisach, J.; Blumberg, W.E. Structural implications derived from the analysis of electron paramagnetic resonance spectra of natural and artificial copper proteins. *Arch. Biochem. Biophys.* **1974**, *165*, 691–708. [CrossRef]

Publisher's Note: MDPI stays neutral with regard to jurisdictional claims in published maps and institutional affiliations.

coatings

MDPI

Article

Plasma Treatment of Thermally Modified and Unmodified Norway Spruce Wood by Diffuse Coplanar Surface Barrier Discharge

Zuzana Košelová, Jozef Ráheľ and Oleksandr Galmiz *

Department of Physical Electronics, Masaryk University, Kotlarska 2, 61137 Brno, Czech Republic; zuzanakoselova@gmail.com (Z.K.); rahel@mail.muni.cz (J.R.)
* Correspondence: oleksandr.galmiz@gmail.com

Abstract: This work deals with the treatment of wood surfaces by diffuse coplanar surface barrier discharge (DCSBD) generated at atmospheric pressure. The effect of the distance of the sample from the electrode surface and the composition of the working gas in the chamber was studied. Norway spruce (*Picea abies*) wood, both unmodified and thermally modified, was chosen as the investigated material. The change in the surface free energy (SFE) of the wood surface was investigated by contact angles measurements. Chemical and structural changes were studied using infrared spectroscopy, X-ray photoelectron spectroscopy (XPS), and scanning electron microscopy (SEM). Activation at a 0.15 mm gap from the electrode led in all cases to an increase in the SFE. The largest change in SFE components was recorded for wood thermally modified to 200 °C. At a 1 mm gap from the electrode increase of SFE occurred only when oxygen (O_2) and argon (Ar) were used as working gas. Treatment in air and nitrogen (N_2) resulted in an anomalous reduction of SFE. With the growing temperature of thermal modification, this hydrophobization effect became less pronounced. The results point out the importance of precise position control during the DCSBD mediated plasma treatment. A slight reduction of SFE on thermally modified spruce was achieved also by short term ultra-violet (UV) light exposure, generated by DCSBD.

Keywords: Norway spruce; thermally treated wood; DCSBD; plasma treatment; surface free energy

Citation: Košelová, Z.; Ráheľ, J.; Galmiz, O. Plasma Treatment of Thermally Modified and Unmodified Norway Spruce Wood by Diffuse Coplanar Surface Barrier Discharge. *Coatings* **2021**, *11*, 40. https://doi.org/10.3390/coatings11010040

Received: 3 December 2020
Accepted: 27 December 2020
Published: 1 January 2021

Publisher's Note: MDPI stays neutral with regard to jurisdictional claims in published maps and institutional affiliations.

1. Introduction

Wood is a well-known industrial material, and its surface properties have a huge impact on durability, quality, and product aesthetics. If the wood has been properly processed and treated, it will last for a very long time, which can be demonstrated on historic buildings, utility and artistic objects, musical instruments, and other wood products. By modifying wood surfaces, it is possible to achieve even better properties, streamline its use, or find new possibilities for its application in various areas of human activity [1].

Plasma treatment can change the surface properties of the wood and have a positive effect e.g., when applying varnishes and adhesives. This could not only affect the amount of the used paint or adhesives but also obtain a higher strength of the glued joints in stressed places.

Diffuse coplanar surface barrier discharge (DCSBD) enables homogeneous surface treatment of various flat surface materials such as wood [2,3], polymers [4], or glass [5]. Concerning the treatment of lignocellulose materials, it was observed that the effect of DCSBD treatment depends on the distance between the treated wood surface and the DCSBD electrode [6,7].

Understanding the reasons behind this effect could help us determine appropriate conditions for plasma treatment and maximize the desired effect, which may find its use in the construction industry, the production of furniture, musical instruments, decorative objects, etc.

In this research a spruce wood was chosen for treatment, owing to its large abundance in Central Europe, mainly due to the recent bark beetle calamity. Nowadays, the spruce wood is frequently thermally modified, usually in the range of 160 °C to 220 °C at low oxygen environment, to improve its dimensional stability against the moisture content and biological resistance against decay [8]. The heat treatment results in the loss of hemicelluloses mainly, which causes the growth in the relative abundance of hydrophobic lignin. This is considered to be the main reason behind the obtained reduced hygroscopicity (ability to absorb and retain water moisture) of wood. At the same time, however, the low surface wettability of thermally modified wood (TMW) complicates its further processing when applying water-based varnishes and adhesives or wood preservatives. For instance, in [9] authors observed a considerable, almost 1/3 reduction in adhesion strength of alkyd-reinforced acrylate paint with TMW spruce, already after a mild treatment (<200 °C).

TMW exhibits higher content of hydrophobic lignin, which is known to be quite sensitive to plasma treatment [10]. Thus, properly applied plasma treatment can be used to revert the obtained hydrophobic surface characteristics of heat-modified spruce as well as other wood species, such as beech, pine core, and other woods with higher lignin content.

In this research, spruce with different heat pre-treatment (160, 180, and 200 °C) was studied. The effect of DCSBD plasma treatment was evaluated by measuring the changes in wood surface free energy (SFE), chemical composition, and micro-morphology. An important parameter that affects the plasma treatment of wood is the composition of plasma treatment working gas. For this sake, the most common industrial gases such as air, nitrogen (N_2), oxygen (O_2), and argon (Ar) were investigated. Finally, factors such as substrate heating and ultra-violet (UV) radiation on the resulting hydrophilicity of the material were studied.

2. Materials and Methods

2.1. Samples and Wood Thermal Modification

The spruce samples were cut into regular cuboids. Because wood is a biological material that is exposed to varying different conditions during its growth, the resulting material is inhomogeneous. It has different surface and internal properties in different places. Only smooth parts of the samples without visible defects were used for the measurement. The samples were stored at room temperature in a closed container that maintained a constant 35–37% humidity without direct sun contact. The moisture content of the samples was measured to be less than 3%. The density of spruce and spruce heat-treated at 160, 180 and 200 °C was measured to be 480, 440, 410 and 400 kg/m^3, respectively. The size of samples varied depending on the analytical technique that was used. Usually, at least 3 samples for each parameter were measured. The Katres s.r.o pilot plant laboratory chamber was used to perform thermal modification of the wood, which thermally treats the wood in the water vapor environment. The process has three basic phases: high-temperature drying, thermal modification, and final cooling. First, the temperature in the chamber, due to hot steam, rises rapidly to 100 °C. It then increases further, but more slowly, to 130 °C. During this, the humidity in the wood drops to a close to zero value. After drying, thermal modification follows. In our case, it took place for three hours at temperatures of 160, 180 or 200 °C. Steam is used during drying and heat treatment as protection against ignition. Finally, the chamber is cooled to a temperature of about 90 °C. At this stage, wetting is important for the final properties of the wood to be usable, it should have a moisture content above 4% [11].

2.2. Plasma Treatment

DCSBD was used for plasma treatment (Figure 1). The electrode system consisted of 32 parallel silver electrodes, which were 1.5 mm wide, 220 mm long, and had gaps of 1 mm between them As the dielectric, 96% alumina ceramics were used (Figure 2). For the plasma activation in a certain gaseous atmosphere, a DCSBD in a closed vessel with holes for the gas outlet, and an inlet with a controlled current flow was used. Then the specific gas for 5 min at a flow rate of 2 L/min was filled, so that the chamber was flushed many times with the given gas and the presence of atmospheric gases was minimal. The sinusoidal voltage was in the amplitude of 10 kV at a frequency of 15 kHz. The generator output power was 400 W for all gases except Ar, where it was 250 W. The efficiency of the whole system is approximately 90%. The distance from the sample to the electrode was controlled with the glass plates (0.15 and 1 mm) positioned on the electrode.

Figure 1. Photo of the DCSBD plasma reactor.

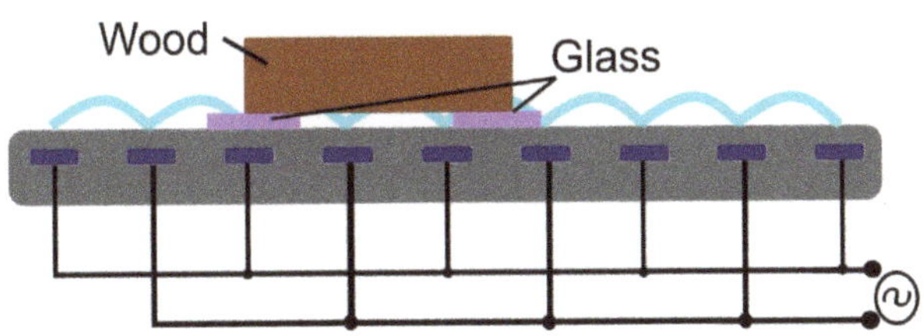

Figure 2. Schematic setup of the DCSBD electrode.

2.3. *Surface Characterization*

The indirect determination of SFE and its polar and dispersive components was done using the Owens-Wendt regression method described previously in more detail [12]. Four liquids were used in this study: distilled water (γ^D = 21.9 mJ/mm^2, γ^P = 51.0 mJ/mm^2), ethylene glycol (γ^D = 29.0 mJ/mm^2, γ^P = 19.0 mJ/mm^2), diiodomethane (γ^D = 50.8 mJ/mm^2, γ^P = 0 mJ/mm^2) and glycerol (γ^D = 28.3 mJ/mm^2, γ^P = 36.9 mJ/mm^2) [13]. Surface Energy Evaluation System (Advex Instruments, Brno-Komín, Czech Republic) was used to measure contact angles (CA) directly from the camera images. For each testing liquid, the contact angle of 15 droplets (1 μL) was measured and the average values were used for the Owens-Wendt regression. The contact angles were determined at the time when the wetting rate becomes constant (dθ/dt = const) [14]. The obtained data were analyzed by Analysis ToolPak of MS Excel 2016 software (Microsoft Corp., Redmond, WA, USA). The normality of the data distribution was verified by its descriptive statistics tool. The significance of differences among the results was tested using the Student's *t*-test, with the significance level of rejecting the null hypothesis being equal to 0.05. The surface morphology of the examined spruce was studied with a scanning electron microscope (SEM) MIRA3 from TESCAN (Brno, Czech Republic). Samples were cut into small pieces of about 15 × 10 × 5 mm^3. Before SEM imaging, the samples were coated with a 10-nm Au–Pd composite layer. The sample surface was electrically connected with the sample holder to reduce surface charge accumulation. All measurements were taken using a secondary electron detector with an accelerating voltage of 7 kV to ensure minimal damage to the surface. The focus was mainly on the internal structures of wood, its tissues, vascular bundles, and especially on places where the effect of plasma etching was evident.

X-ray photoelectron spectroscopy (XPS) measurements were carried out on an ESCALAB 250Xi (Thermo Fisher Scientific, East Grinstead, UK). An X-ray beam with a power of 200 W (650 μm^2 spot size) was used. The survey spectra were acquired with a pass energy of 50 eV and a resolution of 1 eV. High-resolution scans were acquired with a pass energy of 20 eV and a resolution of 0.1 eV. To compensate for the charges on the surface, an electron flood gun was used. Spectra were referenced to the hydrocarbon type C1s component set at a binding energy of 284.8 eV. Spectra calibration, processing, and fitting routines were done using Avantage software.

Attenuated total reflectance (ATR) infrared spectra were measured with Bruker Vertex 80 V spectrometer (Optik Instruments s.r.o., Brno, Czech Republic) utilizing a diamond crystal for ATR. All measurements were taken in an evacuated regime at a maximum pressure of 5 hPa. The spectra were acquired in a range of 4000–800 cm^{-1} with a resolution of 4 cm^{-1}. It was confirmed that the absorption of the band in the region 1031–1053 cm^{-1} did not change as a result of plasma treatment, and each spectrum was normalized to the intensity at 1024 cm^{-1}. At least, four samples of each parameter with a minimum of 3 points on each sample were measured. The results presented are the average of the obtained data.

2.4. Thermal Camera

For temperature measurement, a non-contact method using a thermal camera was taken. This method relies on electromagnetic radiation from objects. For the accuracy of measurement, it is essential to choose the value of emissivity correctly. During the recording we used e = 0.86 (mentioned e.g., in [15]). Emissivity in the range e = 0.82–0.89 is reported in the sources (e.g., [16,17]).

3. Results

3.1. Thermally Modified Wood

Spruce wood samples which were modified with three different temperatures before the plasma treatment—160, 180, and 200 °C are hereafter referred to as T160, T180, T200, respectively. Wood that has not been thermally modified is referred to as a reference—Ref. The thermal modification of timber caused a decrease in SFE. From the initial value of 66 mJ/mm^2, it decreased to 62, 65, and 58 mJ/mm^2 for T160, T180, and T200, respectively. The reference sample has the highest SFE as shown in Figure 3, i.e., wood thermal modification increases the wood surface hydrophobicity. The reason behind this is the intended disruption of cellulose, hemicellulose, and to a lesser extent of lignin, which is more stable. The hydroxyl groups present in hemicelluloses are strongly involved in wetting the wood by forming hydrogen bonds with water molecules. Their decomposition gives rise to observed hydrophobicity growth [18].

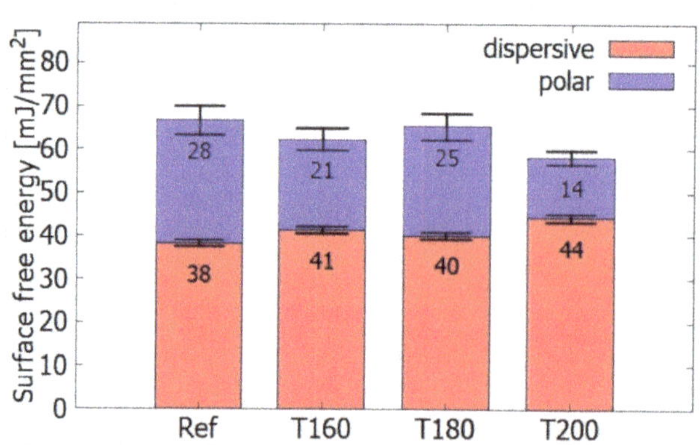

Figure 3. Comparison of the surface energy of different TMW without plasma treatment.

The loss of these hydrogen bonds will also affect the ratio of surface energy components. The dispersion component increases from 38 mJ/mm^2 for the reference wood to 44 mJ/mm^2 for the T200. The polar component decreases from 28 to 14 mJ/mm^2. There are boundaries between T180 and T200 where the hydrophilic behavior is significantly weakened. This is in agreement with an increase in the proportion of lignin in the wood. According to the article [10], there is a small difference between woods heated for 2 h at 150 and 180 °C—28%, compared to 31% for 200 °C.

3.2. Distance from the Electrode

Plasma treatment at the power of 400 W was done for two different distances between the wood and electrode of 0.15 mm and 1 mm. DCSBD plasma is most intense at the height interval of 0.1–0.3 mm from the electrode. When the sample is exposed to DCSBD plasma at this distance range, the contact angles of the water is expected to drop most significantly, as the SFE grows. In our case, however, when treating the reference spruce samples at 0.15 mm, there was no statistically significant change in SFE (Figure 4). The reason is that untreated spruce wood has already relatively high SFE. The resulting low contact angle (especially for water) interferes with the measurement accuracy. As a matter of fact, a value of 70 mJ/mm^2 could represent an upper limit of reliably measurable SFE, and thus any improvement above this level (e.g., by plasma treatment) is undetectable. For thermally treated samples (Figure 4), where the initial SFE was lower, treatment at 0.15 mm resulted in a statistically significant increase of SFE, chiefly due to the growth of its polar component.

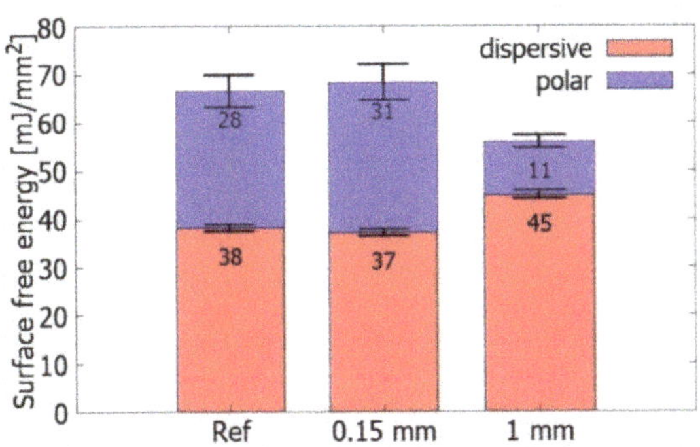

Figure 4. Comparison of surface energies of thermally unmodified wood at different distances from the electrode in the air atmosphere.

The picture changed for the gap of 1 mm. At this position, the opposite effect occurred—the SFE dropped from 66 mJ/mm^2 to 56 mJ/mm^2, on account of polar from 28 to 11 mJ/mm^2, although dispersive component manifested a slight increase from 38 to 45 mJ/mm^2 (Figure 4). In article [7], the authors also reported such anomalous increase in hydrophobicity for the distance of 0.93 mm, where plasma began to quench.

The hydrophobization effect of 1mm distance treatment was observed also for thermally modified samples (Figure 5). The most significant SFE increase occurred for T200 at a distance of 0.15 mm, which is consistent with the results from [19]. This is related to the lower initial SFE of wood T200. After plasma treatment, the total energy was comparable to other samples with different heat treatments. The total values of the SFE of all thermally modified and reference wood for the distance of 0.15 mm from the electrode were in the range of 37–38 mJ/mm^2 for dispersive and 27 to 31 mJ/mm^2 for polar components. For a distance of 1 mm, the dispersion components were in the range of 45–46 mJ/mm^2 and the polar components were in the range of 9–11 mJ/mm^2.

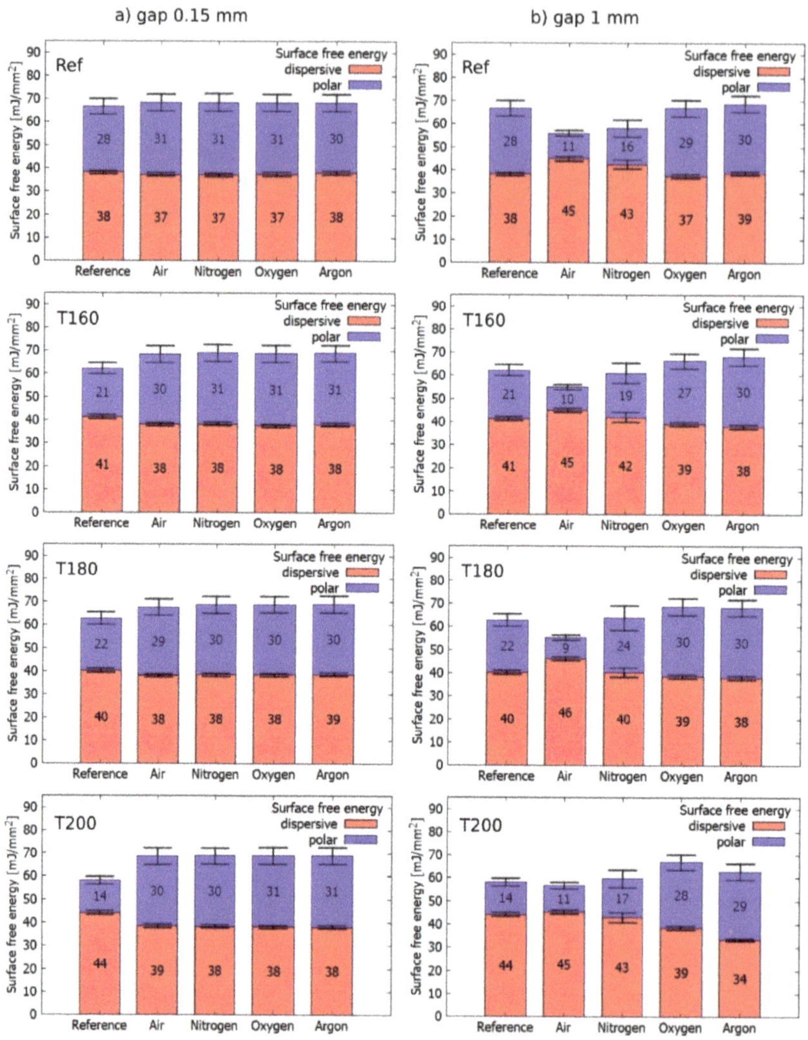

Figure 5. Comparison of surface free energies for different gases during plasma activation of wood. (**a**) in case of 0.15 mm gap and (**b**) in case of 1 mm gap from the electrode.

3.3. *Influence of Gas Composition on Processing*

As a working gas, gases commonly used in industry, i.e., N_2, O_2, Ar, and air were studied. The wood treatment took place in each atmosphere at two different distances of the material from the electrode (0.15 and 1 mm). At the distance of 0.15 mm from the electrode, the SFE increases slightly in all atmospheres. In particular, the dispersion component slightly decreases, and the polar component increases. This can be explained by oxidation on the surface either directly during plasma activation or after extraction into atmospheric air, where the interaction of activated wood with O_2 can take place. So, it could be concluded that while keeping the gap between the sample and the plasma in the range of 0.1–0.3 mm the working gas does not influence significantly the plasma treatment effect. In practice, however, it is hard to keep such a high precision during wood processing.

On the contrary, in the case of a 1 mm distance, the working gas played an important role. Air and N_2 treatment displayed a more hydrophobic wood surface. With this respect, air gas showed a more pronounced effect. For both gases, the dispersion component increased, and the polar component decreased. Interestingly enough, 1 mm distance treatment in pure O_2 had no hydrophobization effect. This points out the important role of NO_x chemistry in the effect. An ample amount of nitric oxide compounds (NO, NO_2, N_2O_5) and nitric acid in the presence of H_2O is formed within the DCSBD and can strongly interact with the surface. These compounds oxidize the surface, thus rather causing an increase in SFE [20,21]. At the moment, it is an unresolved question to what extent one can manipulate the resulting SFE by altering the N_2 to O_2 ratio. This question was partially addressed in [22], where authors tested different atmospheres for DBD ATMOS plasma activation and monitored the change in wetting. They got an increase in the contact angle for 1:2 and 3:1 ratios N_2:O_2 and a decrease for the case 1:1. However, they did not explain this behavior but only stated that "the combination of structural change (induced by UV radiation, the impact of metastable particles, or both) and chemical change due to surface oxidation are responsible for the observed surface modification of wood samples [22].

In the case of O_2 and Ar, the total SFE increased, comparable to the case where the wood was treated at a distance of 0.15 mm from the electrode. The increase of the polar component was slightly larger (max. 4 mJ/mm^2, but mostly up to 1 mJ/mm^2) and in most cases, the decrease of the dispersion component was slightly smaller (max. 5 mJ/mm^2 for T200 Ar, but most up to 1 mJ/mm^2). The behavior of T200 wood was most different from other wood samples.

The polar component in the N_2 atmosphere after treatment at 1 mm distance increased with a temperature of thermal modification, but for T200 fell again close to the reference wood (Ref 1 mm) value.

3.4. Comparison of Wood and Polymer

To verify that the increase in water contact angle at 1 mm is specific to wood chemical composition, the contact angle of water for the polymer treated was measured. Specifically, polymethyl methacrylate (PMMA) commonly known as plexiglass was used. After plasma treatment, the hydrophilicity of the plexiglass increased, see Figure 6. For PMMA, the water contact angle was reduced for both cases. Keeping a 0.15 mm gap, the water contact angle decreased more markedly than for a 1 mm gap. Wood, on the other hand, acquired more hydrophilic properties only for 0.15 mm, and, at a distance of 1 mm, it was hydrophobized. Therefore, it is a matter of wood material and its specific morphological and chemical composition. For some polymers, plasma treatment can also increase the hydrophobicity of the surface due to etching, the formation of nanostructures and can achieve even superhydrophobic behavior [23]. However, in Section 3.7.2 Wood etching it will be shown that the increase in hydrophobicity in the wood was not caused by emerging nanostructures, as in these cases.

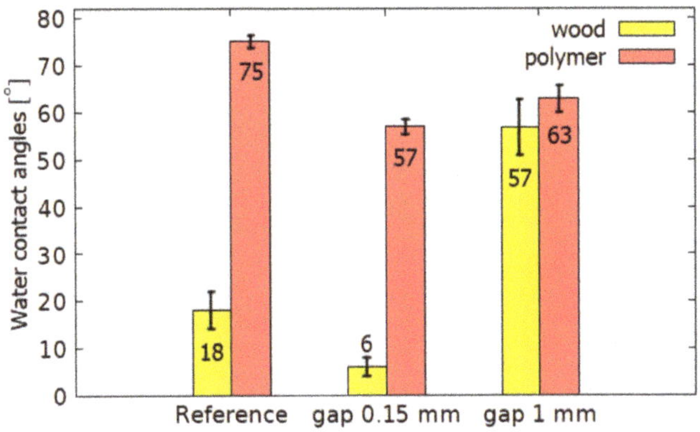

Figure 6. Comparison of water contact angles between polymer and wood.

3.5. Fourier Transform Infrared Spectroscopy

FTIR spectra for T200 and both distances treatment are shown in Figure 7. There was no observable change in the absorbance band of the bonds of water and OH groups (3550–3150 cm^{-1}). The absorbance in the CH_x band (2947 cm^{-1}) decreased after plasma treatment. This decrease is due to chemical reactions on the sample surface [2]. In the area of C=O conjugate bonds (1655 cm^{-1}) there was a significant increase in absorbance for samples treated at 0.15 mm compared to the reference sample. These bonds are affected by plasma oxidation [24]. In [24] for spruce, both regions corresponding to C=O increased, of which the unconjugated band was more significant. In the case of a 1 mm gap, this peak decreased. Unconjugated bonds (1727 cm^{-1}) increased after plasma treatment in both cases.

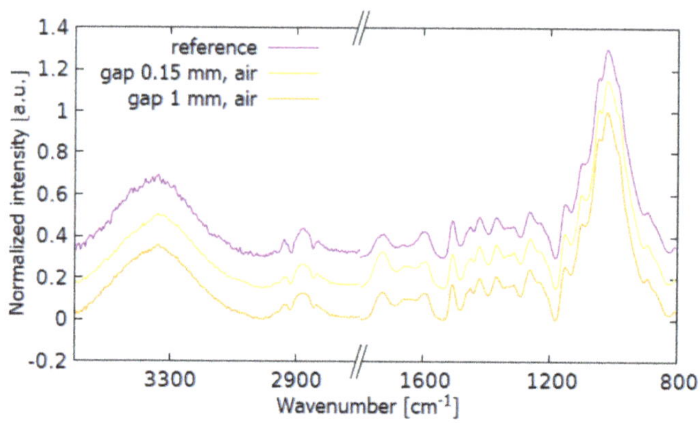

Figure 7. FTIR spectra of wood T200 to compare the change in chemical composition at different distances from the electrode. The treatment was done in the air.

The 1596 cm^{-1} peak decreased for 0.15 mm that supports the theory that in the active plasma region CH_x components undergo degradation. At the same time after plasma treatment at a 1 mm distance, this peak decreased less. For plasma-treated samples, absorbance increases in the syringyl and guaiacyl regions (aromatic groups from lignin) and OH groups from cellulose (1313–1336 cm^{-1}).

Figure 8 compares measurements in N_2 and air. The samples behave very similarly. For T180, the band of unconjugated C=O after the plasma treatment in the air hardly changes. At a distance of 1 mm they are almost identical, except for a large increase in C=O (1727, 1655 cm^{-1}) and an increase in vibrations of 1264, 1219 cm^{-1} also bound to C=O bonds. These differences are also evident in the treatment at a distance of 0.15 mm from the electrode, especially in the vicinity of the areas 1727 and 1655 cm^{-1}. Greater oxidation occurred in the N_2 atmosphere.

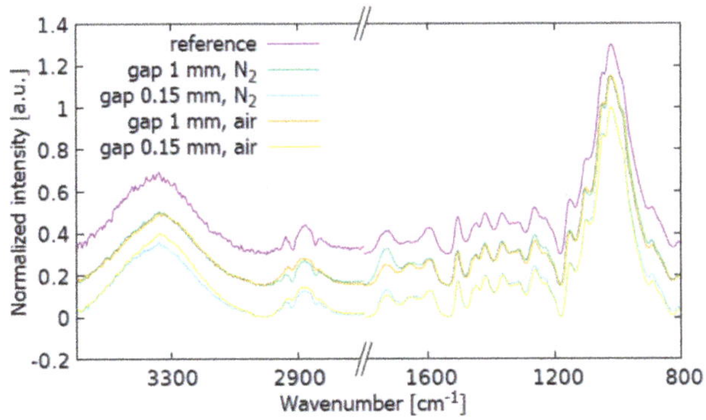

Figure 8. FTIR spectra comparison of the T180 wood after plasma treatment under N_2 and air atmosphere.

3.6. X-ray Photoelectron Spectroscopy

Due to the complex irregular surface topography, the random distribution of the wood components, and their random size, the errors of some measurements are significant.

Some samples contained a small percentage of N_2 (up to 1.5%) in addition to O_2 and carbon. After plasma treatment in a N_2 atmosphere, they reached a maximum of 2% at 1 mm distance from the electrode and 9% at 0.15 mm distance. This increase did not occur in the air. The oxygen and carbon atomic percentages and O/C ratio obtained from XPS spectra for various gases and distances from the electrode are presented in Table 1.

Table 1. Oxygen and carbon atomic percentage and O/C ratio obtained from XPS spectra for various gases and distances from the electrode. Data are in [%], except for O/C, which is dimensionless.

		Ref	Air		N_2		Ar		O_2
			0.15 mm	1 mm	0.15 mm	1 mm	0.15 mm	1 mm	1 mm
Ref	O	28 ± 2	38 ± 2	30 ± 4	37 ± 1	28 ± 5	34 ± 3	29 ± 7	32 ± 3
	C	72 ± 1	62 ± 3	69 ± 4	57 ± 1	70 ± 1	66 ± 7	71 ± 7	68 ± 3
	N	-	-	-	6 ± 1	2 ± 1	-	-	-
	O/C	0.38(2)	0.62(5)	0.44(6)	0.65(2)	0.40(9)	0.53(5)	0.41(6)	0.46(4)
T180	O	29 ± 2	34 ± 1	29 ± 1	-	26 ± 1	-	28 ± 2	-
	C	71 ± 4	66 ± 1	71 ± 1	-	74 ± 1	-	72 ± 2	-
	O/C	0.41(4)	0.51(2)	0.40(1)	-	0.35(1)	-	0.39(2)	-
T200	O	29 ± 2	40 ± 1	26 ± 3	32 ± 1	-	33 ± 1	-	34 ± 1
	C	71 ± 2	60 ± 1	73 ± 4	59 ± 2	-	63 ± 2	-	66 ± 2
	N	-	-	-	9 ± 2	-	4 ± 1	-	-
	O/C	0.41(3)	0.66(1)	0.34(5)	0.54(3)	-	0.52(2)	-	0.52(2)

Component C1 corresponds to bonds C–C and C–H, component C2 is attributed to bonds C–O, component C3 can correspond to either the group O–C–O or the bond C=O and component C4 corresponds to the groups O=C–O. The main contribution to peak C1 comes from lignin and extracts, in peaks C2 and C3 it comes from functional groups in lignin and polysaccharide. The C4 peak is attributed to hemicellulose [19].

It is not surprising that after plasma treatment of a sample, the oxidation on its surface increased when it was directly in contact with the plasma. There was also a decrease in the Cl peak due to the decomposition of the extracts. This decrease was more pronounced in the air than in the N_2 atmosphere. There was no measurable change in Ar. An increase in C3 and C4 peaks was also observed. A graphical representation of component changes can be seen in Figure 9a,b, where the drop of the C1 component and the increase of the C3, C4 components are clearly shown. It is known that an increase in the polar part of SFE occurs due to oxidation [19].

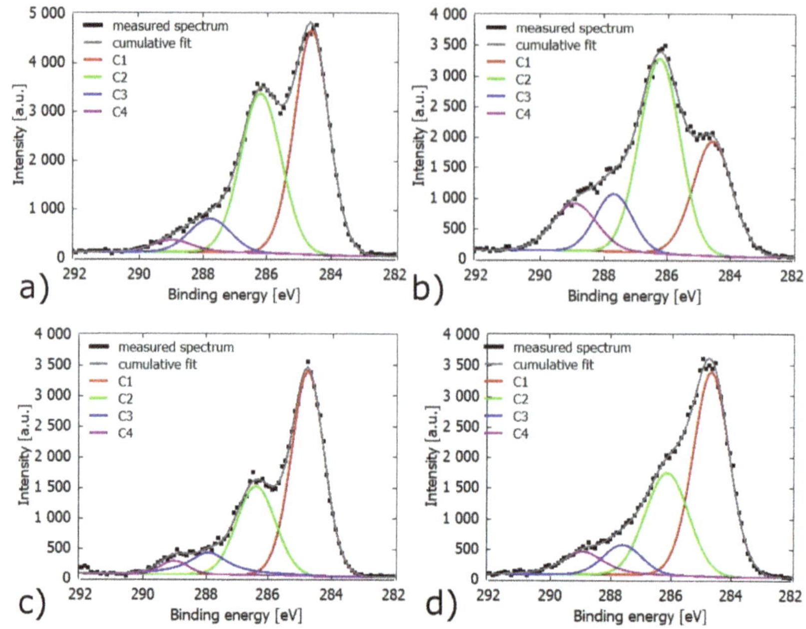

Figure 9. The XPS spectrum deconvolution of Cls band: (**a**) untreated Ref sample, (**b**) Ref sample plasma treated in the air at 0.15 mm gap (**c**) Ref sample plasma treated in the air at 1 mm gap, (**d**) T200 plasma treated at 1 mm gap.

In the case of a 1 mm gap under a N_2 and air atmosphere, the obtained O/C and heights of Cl peaks values were similar to the reference sample. A decrease in the O/C ratio on TMW was recorded. This corresponds to the increase in SFE measured by the SeeSystem. Figure 9c,d show examples of XPS spectra of reference and T200 samples with the corresponding individual components. It is seen that even after plasma treatment, a similar ratio of peaks between the reference and heat-treated wood was obtained.

For a 1 mm gap, the increase in O/C was greater for O_2, thanks to the easiest oxidation. There was also an obvious decrease in the Cl component, from 48% to 17% for the reference wood and from 46% to 29% for T200.

At a 0.15 mm gap from the electrode for Ar, N_2, and air atmospheres a significant increase in O_2 values was recorded. At a 1 mm gap, lower O/C values for TMW after plasma treatment were obtained compared to the reference ones, though they were within the error bars.

3.7. Other Influences

3.7.1. Wood Heating

The samples are heated during the plasma treatment. The resulting temperature was determined using a thermal camera. The error of measurement was 2 °C.

Detection photos were taken right after removal from the DCSBD on the side that was placed into the plasma region. It was noticed that the wood did not heat up evenly, that is because the wood is not perfectly smooth. After 10 s, the maximum temperature of 52.2 °C was reached for the 0.15 mm gap. At a greater distance from the plasma—1 mm—it was only 40.8 °C.

The measurement of the temperature in the case of O_2 and Ar atmosphere was affected by the procedure of getting the samples out of the reactor. It is assumed that the maximal temperature values were similar to the values obtained in the case of air treatment.

To separate the influence of the temperature, the untreated sample was heated to 60 °C by a hot air gun and measured with SeeSystem and SEM. The result did not show observable changes in surface properties. The SFE changed by a maximum of 2.5% of the total value. In Table 2 the highest achieved a temperature of activated wood under different conditions could be seen.

Table 2. The maximum measured temperature of the wood sample surface.

Distance.	0.15 mm Gap			1 mm Gap		
Working gas	O_2	Ar	Air	O_2	Ar	Air
Temperature (°C)	49.5	38.2	52.2	38.5	35.6	40.8

3.7.2. Wood Etching

The structural changes of the sample surface after the plasma treatment were monitored using SEM. No significant changes in the structures were observed after the thermal nor plasma modification (Figures 10 and 11).

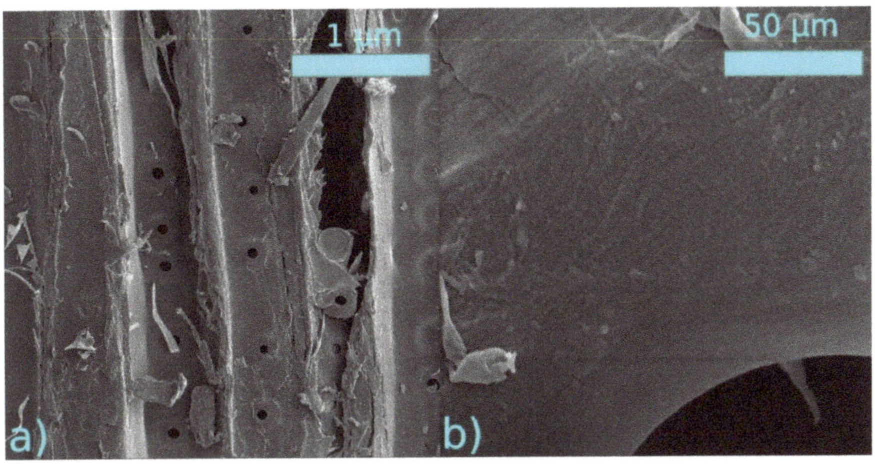

Figure 10. SEM images of plasma-treated samples at a distance of 0.15 mm from the electrode (**a**) 1000× magnification, (**b**) 50,000× magnification.

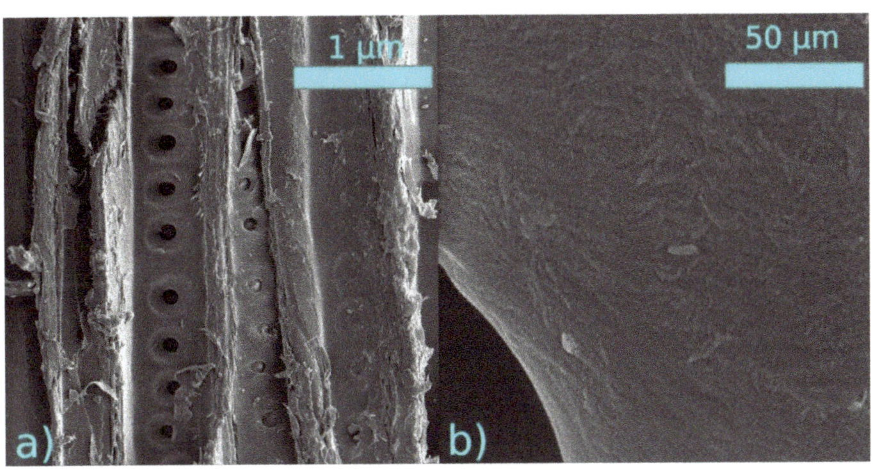

Figure 11. SEM images of plasma treated samples at a distance of 1 mm from the electrode (**a**) 1000× magnification, (**b**) 50,000× magnification.

Previously it has been reported [25], that DCSBD treatment can etch a wood surface. Though the time needed for such a modification is in order of minutes. In [3], researchers laid wood directly on the DCSBD and did not observe the formation of nanostructures even after 60 s.

3.7.3. Influence of UV Radiation

To evaluate the sole effect of UV radiation emitted by plasma discharge a quartz glass was placed between the plasma and the wood. This allowed transmission of UV radiation, but at the same time preventing the passage of ions, electrons, radicals, and other molecules from the plasma, such as ozone or nitrogen oxides. The sample was positioned 1 mm above the electrode, and the air was used as the working gas.

Figure 12a,b show the resulting total SFEs and their corresponding disperse (dis.) and polar (pol.) components. Both plasma and UV treatment resulted in a statistically significant reduction of SFE. In both cases, the reduction could be attributed to the lowering of SFE's polar component, which was more pronounced for plasma treatment. The sensitivity of thermally modified spruce to UV exposure contradicts our former observation in [26], where thermally modified European beech (Fagus sylvatica) did not exhibit any statistically significant response. The reason for this remains unclear yet. One explanation can be higher initial SFE of spruce (58 vs. 49 mJ/mm^2 for T200), which may allow the effect to manifest itself. Another explanation may lay in the different chemical composition of coniferous spruce and deciduous beech.

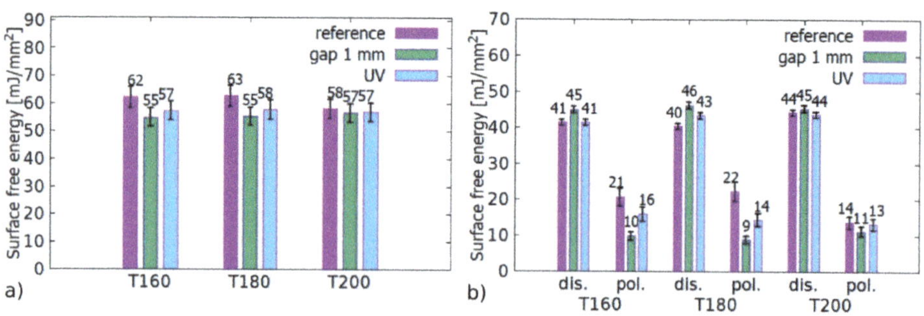

Figure 12. Comparison of surface energies of reference samples, plasma treated at 1 mm gap and treated by plasma generated UV. (**a**) Total SFE, (**b**) division into SFE components.

To better understand this discrepancy, further thorough testing with different wood species and UV intensity would be needed. This is beyond the scope of this article and can be considered for further studies. UV radiation has been reported to damage lignin and cellulose crystallinity upon prolonged exposure [27]. However, during the presented experiments, the UV expose of samples was for a short time. The effect of UV radiation on color was studied in articles [28,29] and FTIR tests were performed. They observed darkening of the wood in the reference samples and lightening of the heat modified samples.

4. Discussion

Thermal modification of wood should raise the relative content of hydrophobic lignin. The results of SFE measurements and the O/C ratio of TMW (Figure 3, Table 1) confirmed this expectation. A large jump of 4.6 mJ/mm^2 in SFE occurred between T180 and T200. The O/C ratio for TMW gave significantly higher values for T160, while there was only a slight decrease between T180 and T200. However, the measurement on sample T160 is suspicious due to the lower proportion of the Cl component compared to other plasma untreated samples. Deviation from other measurements may be due to sample inhomogeneity. According to [19,30], one would expect thermal modification to reduce the O/C ratio. From the XPS measurements, it was observed that the components C2, C3, C4 did not change significantly (except for T160 wood). The Cl component slightly decreased compared to the reference wood, although an increase due to its association with lignin was expected. However, within the measurement accuracy, values remain the same as do the O/C ratios of Ref, T180, T200.

Only for the T200 sample the FTIR measurements (Figure 7) of showed an unambiguous increase in the absorbance regions of the aromatic skeletal bonds in lignin and the C=O bonds (1596, 1655, 1727 cm^{-1}).

The SFE of the samples treated at a 0.15 mm gap from the electrode showed some increase for all types of the atmosphere (Figure 5a). In general, the starting SFE difference of TMW made at different temperatures was lost after the 0.15mm plasma treatment. XPS data showed a large increase in the O/C ratio and an increase in components C2, C3, and C4 at the expense of C1. This suggests a higher presence of polar functional groups on the surface. These results, together with the increase in C=O bonds from FTIR measurements (Figures 7 and 8) signify surface oxidation. The XPS data indicate greater oxidation of pristine spruce for air and N_2 and lower for Ar. Plasma treatment of TMW in Ar and N_2 showed similar O/C ratio, as well as values of SFE. This may be due to a different process of increasing SFE. FTIR shows a larger increase in C=O under N_2 compared to air (Figure 8). The surface energies obtained from laid liquid droplets are in accordance with these FTIR measurements, i.e., higher in a N_2 atmosphere than in air.

Zanini et al. studied chemical changes in wood after treatment with Ar RF plasma [31]. An increase in the concentration of phenoxy radicals was observed. The formation of radicals occurs first on lignin, which becomes significantly modified. The resulting radicals can then react with other monomers to form C–C or C–O bonds. However, the RF plasma used in [31] has a distinct set of discharge operation conditions to DCSBD, thus one should expect that both types of plasma would act differently on wood chemical bonds.

For a 1 mm gap, the surface energy strongly depended on the used working gas (Figure 5b). DCSBD plasma in air forms reactive intermediates such as O_2^*, O_2, O_3, O, O^+, O_3^+, e^-, OH, N, CO_2, N_2^* [6], but only long-lived neutrals and UV is capable to reach the sample surface positioned at 1mm distance. For air, this would be ozone or nitrogen oxides. Considering the frequency range 15–50 kHz AC, 15 kHz is the most suitable for generating O_3 and NO [32].

Samples treated in the air at a 1 mm gap showed a reduction in the polar component of the SFE. Changes in values obtained from XPS showed no statistical difference. FTIR measurements showed a small increase in the proportion of unconjugated C=O bonds and a decrease in aromatic skeletal bonds in lignin.

It has been shown that treatment at the 1 mm gap increases the wettability of the cellulose slightly and therefore cannot cause an increase in the hydrophobicity of the wood surface [26]. Not only chemical interaction and UV radiation could cause the change of morphology but the closing of micropores could also be a reason [33]. Partial closure of surface pores has been observed during ionic irradiation of wood [34]. In our conditions, at a 1 mm gap and a treatment time of 10 s, ion radiation should not be significant [22]. The morphology of the wood surface using SEM was mapped. It was found that 10 s is too short a time to etch or otherwise affect the wood morphology. Some morphological changes occur only after a longer treatment period.

The effect of hydrophobization after plasma treatment was not observed on PMMA material after treatment under the same conditions. The effect is therefore associated with the wood material itself, and its chemical changes.

Isolation of the temperature effect has shown that it had no influence on SFE or morphology. Again, the treatment time and temperature were too short to cause any changes.

By isolating the effect of UV radiation, it was found out that it contributed to an increase in the dispersion and a decrease in the polar part with an overall decrease in SFE

(Figure 12). By comparing spruce with beech wood, it can be stated that the effect of UV irradiation differs depending on wood type.

For a 1 mm distance treatment, the results in N_2 and air are very similar. Lack of O_2 during treatment may contribute to a slightly lower increase in hydrophobicity. But part of the water vapor desorbed from the sample could provide O_2 molecules needed for the formation of nitrogen oxides and nitric acid. It should be stated, however, that the effects of secondary nitrogen oxides reactions that may take place after samples removed from the chamber could not be eliminated.

Article by Bihani et al. states that wood-meal (in the presence of O_2) binds gaseous molecules NO and NO_2 relatively quickly [35]. According to the research, reactions with NO and NO_2 did not cause a decrease in SFE [35,36]. Wood-meal was modified with nitrogen oxides for de-lignification. In wool or delignified materials, nitrogen dioxide is converted to nitric acid in the presence of O_2. These reactions should not be significantly affected by lignin, nitric acid can then strongly oxidize wood [36]. This should increase the polar component of SFE.

When treating the T200 sample at 1 mm in air, the decrease of the polar part was negligible. After plasma treatment in N_2, O_2, and Ar, the polar part raised (Figure 5b). A possible explanation is that thermal modification at 200 °C had actually achieved a hydrophobicity threshold/maximum for spruce wood. The source for further chemical reactions resulting in SFE decline is depleted for the T200 sample. Taking into account, that thermal treatment results in the decomposition of hemicelluloses mainly, the following hypothesis may be drawn: the hydrophobization effects it due to chemical reactions of long-lived plasma generated particles and hemicellulose on the wood surface. The same conclusion was derived for thermally modified European beech (*Fagus sylvatica*) in [26].

From the previous experiments, it is known that the plasma treatment of the cellulose increases its hydrophilicity. In a paper by Talviste et al. [26], it was reported that plasma treatment by DCSBD at 1 mm gap has the same tendency as the reference sample on the cellulose paper water uptake. It was concluded that the cellulose itself contributes to the increased hydrophilicity.

O_2 treatment at 1mm caused SFE to increase by increasing its polar component. This is due to the strong oxidizing effects of ozone [21,37]. The selectivity of ozone in reactions with lignin and carbohydrates is strongly dependent on the pH of the environment. When activated by plasma, acidic components such as formic acid (HCOOH) and acetic acid (CH_3COOH) are formed, which select the reactions of ozone with lignin [38]. Ozone reacts mainly with unbound bonds, carbonyl, ether, and hydroxyl groups. DCSBD plasma under an O_2 atmosphere forms up to 2000 ppm O_3 [39]. The O/C ratio also indicates strong oxidation. With respect to the hydrophobization effect, the ozone would definitely contribute to hemicellulose degradation. However, this effect could be completely overshadowed by the formation of novel polar groups on the surface.

5. Conclusions

The wood of Norway spruce (*Picea abies*) was activated using DCSBD plasma, including heat-treated samples. Activation proceeded in atmospheric air and frequent working gases used in industry: N_2, O_2, Ar, all at atmospheric pressure. Two activation distances were considered: in the plasma zone and above it. Activation at a 0.15 mm gap from the electrode led in all cases to an increase in the SFE. That means the starting SFE difference of TMW made at different temperatures was lost.

At a distance of 1 mm from the electrode increase of SFE occurred only when O_2 and Ar were used as working gas. However, it was observed that after treatment in air and N_2 the values of SFE decrease due to polar part reduction. This effect is of importance in the case of industrial plasma application where it is hard to keep the working distance while treating timber. Understanding the reasons behind this effect could help us determine appropriate conditions for plasma treatment and maximize the desired effect.

Wood material was compared to PMMA, which does not show a similar effect. It is therefore a material matter of wood and its morphological and chemical composition. After the SEM analyses, it could be stated that no morphological changes take place during short plasma treatment. It was shown that heating of wood during the plasma treatment did not influence the SFE, either.

The hydrophobization effects that were shown are explained by the chemical reactions of long-lived plasma generated particles and hemicellulose on the wood surface.

A slight reduction of SFE on thermally modified spruce was achieved also by short term UV light exposure, generated by DCSBD at a 1 mm distance. From the comparison of spruce with beech wood, it could be stated that the effect of UV irradiation differs depending on wood type. To fully understand the effect of UV on wood, further thorough testing on multiple types of wood could be recommended and UV effects such as short exposure time, intensity, etc. to be assessed.

Author Contributions: Conceptualization, O.G., and J.R.; formal analysis, Z.K.; investigation, Z.K.; writing—original draft preparation, Z.K., and O.G.; writing—review and editing, O.G. and J.R.; supervision, O.G. All authors have read and agreed to the published version of the manuscript.

Funding: This research was funded by the Ministry of Education, Youth, and Sports of the Czech Republic, grant number LM2018097.

Institutional Review Board Statement: Not applicable.

Informed Consent Statement: Not applicable

Data Availability Statement: Data available in a publicly accessible repository. The data presented in this study are openly available in [repository name e.g., FigShare] at [doi], reference number [reference number].

Acknowledgments: We would like to thank the Department of Wood Science at the Mendel University in Brno for providing the thermally treated wood samples.

Conflicts of Interest: The authors declare no conflict of interest.

References

1. Dejmal, A. *Sušení a Modifikace Dřeva (Učební Texty) [Wood drying and modification (Textbook)]*; Mendel University in Brno: Brno, Czech Republic, 2017.
2. Odrášková, M.; Ráhel', J.; Zahoranová, A.; Tiňo, R.; Černák, M. Plasma activation of wood surface by diffuse coplanar surface barrier discharge. *Plasma Chem. Plasma Process.* **2008**, *28*, 203–211. [CrossRef]
3. Gerullis, S.; Kretzschmar, B.S.; Pfuch, A.; Beier, O.; Beyer, M.; Grünler, B. Influence of atmospheric pressure plasma jet and diffuse coplanar surface barrier discharge treatments on wood surface properties: A comparative study. *Plasma Process. Polym.* **2018**. [CrossRef]
4. Kormunda, M.; Homola, T.; Matousek, J.; Kovacik, D.; Cernak, M.; Pavlik, J. Surface analysis of poly(ethylene naphthalate) (PEN) films treated at atmospheric pressure using diffuse coplanar surface barrier discharge in air and in nitrogen. *Polym. Degrad. Stab.* **2012**, *97*, 547–553. [CrossRef]
5. Homola, T.; Matoušek, J.; Kormunda, M.; Wu, L.Y.L.; Černák, M. Plasma treatment of glass surfaces using diffuse coplanar surface barrier discharge in ambient air. *Plasma Chem. Plasma Process.* **2013**, *33*, 881–894. [CrossRef]
6. Jablonsky, M.; Smatko, L.; Botkova, M.; Tino, R.; Sima, J. Modification of wood wettability (European beech) by diffuse coplanar surface barrier discharge plasma. *Surfaces* **2016**, *50*, 41–48.
7. Tino, R.; Smatko, L. Modifying wood surfaces with atmospheric diffuse coplanar surface barrier discharge plasma. *Wood Fiber Sci.* **2014**, *46*, 459–464.
8. van Blokland, J.; Olsson, A.; Oscarsson, J.; Daniel, G.; Adamopoulos, S. Crack formation, strain distribution and fracture surfaces around knots in thermally modified timber loaded in static bending. *Wood Sci. Technol.* **2020**, *54*, 1001–1028. [CrossRef]
9. Altgen, M.; Militz, H. Thermally modified Scots pine and Norway spruce wood as substrate for coating systems. *J. Coat. Technol. Res.* **2017**, *14*, 531–541. [CrossRef]
10. Yildiz, S.; Gezer, E.D.; Yildiz, U.C. Mechanical and chemical behavior of spruce wood modified by heat. *Build. Environ.* **2006**, *41*, 1762–1766. [CrossRef]
11. Horníček, S. *Vliv Tepelné Modifikace na Vybrané Vlastnosti Dřeva Plantážově Pěstovaného Topolu [The Effect of Thermal Modification on Selected Properties of Plantation Grown Poplar Wood]*; Mendel University in Brno: Brno, Czech Republic, 2016.
12. Ninness, B. Estimation of I// Noise. *IEEE Trans. Inf. Theory* **1998**, *44*, 32–46. [CrossRef]
13. Gindl, M.; Sinn, G.; Gindl, W.; Reiter, A.; Tschegg, S. A comparison of different methods to calculate the surface free energy of wood using contact angle measurements. *Colloids Surf.* **2001**, *181*, 279–287. [CrossRef]

14. Nussbaum, R.M. Natural surface inactivation of Scots pine and Norway spruce evaluated by contact angle measurements. *Holz Roh Werkst.* **1999**, *57*, 419–424. [CrossRef]
15. Infrared Emissivity Table | ThermoWorks online. Available online: https://www.thermoworks.com/emissivity-table (accessed on 27 December 2020).
16. Mikron Instrument Company Inc. *Table of Emissivity of Various Surfases. Introduction*; Mikron Instrument Company Inc.: Flurlingen, Switzerland, 2002.
17. Wilson, J.S. *Sensor Technology Handbook*; Elsevier: Oxford, UK, 2005; ISBN 9780750677295.
18. Gérardin, P.; Petrič, M.; Petrissans, M.; Lambert, J.; Ehrhrardt, J.J. Evolution of wood surface free energy after heat treatment. *Polym. Degrad. Stab.* **2007**, *92*, 653–657. [CrossRef]
19. Altgen, D.; Avramidis, G.; Viöl, W.; Mai, C. The effect of air plasma treatment at atmospheric pressure on thermally modified wood surfaces. *Wood Sci. Technol.* **2016**, *50*, 1227–1241. [CrossRef]
20. Tangsathitkulchai, C.; Ngernyen, Y.; Tangsathitkulchai, M. Surface modification and adsorption of eucalyptus wood-based activated carbons: Effects of oxidation treatment, carbon porous structure and activation method. *Korean J. Chem. Eng.* **2009**, *26*, 1341–1352. [CrossRef]
21. Kang, G.; Ni, Y.; van Heiningen, A.R.P.; Zhang, Y. Influence of lignins on the degradation of cellulose during ozone treatment. *J. Wood Chem. Technol.* **1995**, *15*, 413–430. [CrossRef]
22. Busnel, F.; Blanchard, V.; Prégent, J.; Stafford, L.; Riedl, B.; Blanchet, P.; Sarkissian, A. Modification of sugar maple (Acer saccharum) and black spruce (Picea mariana) wood surfaces in a dielectric barrier discharge (DBD) at atmospheric pressure. *J. Adhes. Sci. Technol.* **2010**, *24*, 1401–1413. [CrossRef]
23. Dimitrakellis, P.; Gogolides, E. Hydrophobic and superhydrophobic surfaces fabricated using atmospheric pressure cold plasma technology: A review. *Adv. Colloid Interface Sci.* **2018**, *254*, 1–21. [CrossRef]
24. Lux, C.; Szalay, Z.; Beikircher, W.; Kovacik, D.; Pulker, H.K. Investigation of the plasma effects on wood after activation by diffuse coplanar surface barrier discharge. *Eur. J. Wood Wood Prod.* **2013**, *71*, 539–549. [CrossRef]
25. Galmiz, O.; Talviste, R.; Panáček, R.; Kováčik, D. Cold atmospheric pressure plasma facilitated nano-structuring of thermally modified wood. *Wood Sci. Technol.* **2019**. [CrossRef]
26. Talviste, R.; Galmiz, O.; Stupavská, M.; Ráhel', J. Effect of DCSBD plasma treatment distance on surface characteristics of wood and thermally modified wood. *Wood Sci. Technol.* **2020**, *54*, 651–665. [CrossRef]
27. Patachia, S.; Croitoru, C.; Friedrich, C. Effect of UV exposure on the surface chemistry of wood veneers treated with ionic liquids. *Appl. Surf. Sci.* **2012**, *258*, 6723–6729. [CrossRef]
28. Timar, M.C.; Varodi, A.M.; Gurău, L. Comparative study of photodegradation of six wood species after short-time UV exposure. *Wood Sci. Technol.* **2016**, *50*, 135–163. [CrossRef]
29. Srinivas, K.; Pandey, K.K. Photodegradation of thermally modified wood. *J. Photochem. Photobiol. B Biol.* **2012**, *117*, 140–145. [CrossRef]
30. Nguila Inari, G.; Petrissans, M.; Lambert, J.; Ehrhardt, J.J.; Gérardin, P. XPS characterization of wood chemical composition after heat-treatment. *Surf. Interface Anal.* **2006**, *38*, 1336–1342. [CrossRef]
31. Zanini, S.; Riccardi, C.; Canevali, C.; Orlandi, M.; Zoia, L.; Tolppa, E.L. Modifications of lignocellulosic fibers by Ar plasma treatments in comparison with biological treatments. *Surf. Coat. Technol.* **2005**, *200*, 556–560. [CrossRef]
32. Čech, J.; Brablec, A.; Černák, M.; Puač, N.; Selaković, N.; Petrović, Z.L. Mass spectrometry of diffuse coplanar surface barrier discharge: Influence of discharge frequency and oxygen content in N_2/O_2 mixture. *Eur. Phys. J. D* **2017**, *71*. [CrossRef]
33. Gardner, D.J.; Frazier, C.E.; Christiansen, A.W. Characteristics of the Wood Adhesion Bonding Mechanism Using Hydroxymethyl Resorcinol. In *Proceedings of the Wood Adhesives*; Forest Products Society: San Diego, CA, USA, 2005; pp. 93–97.
34. Ramos, H.J.; Monasterial, J.L.C.; Blantocas, G.Q. Effect of low energy ion beam irradiation on wettability of narra (Pterocarpus indicus) wood chips. In *Proceedings of the Nuclear Instruments and Methods in Physics Research, Section B: Beam Interactions with Materials and Atoms*; Elsevier: Pacific Grove, CA, USA, 2006; Volume 242, pp. 41–44.
35. Bihani, B.; Samuelson, O. Consumption of nitrogen oxides during pretreatment of wood with nitrogen dioxide and oxygen. *Wood Sci. Technol.* **1984**, *18*, 295–306. [CrossRef]
36. Subramanian, R.V.; Balaba, W.M.; Somasekharan, K.N. Surface Modification of Wood Using Nitric Acid. *J. Adhes.* **1982**, *14*, 295–304. [CrossRef]
37. Mbachu, R.A.D.; Manley, R.S.J. Degradation of lignin by ozone—1. The kinetics of lignin degradation of ozone. *J. Polym. Sci. Polym. Chem. Ed.* **1981**, *19*, 2053–2063. [CrossRef]
38. Kureková, G. *Wood Surface Treatment by DCSBD Plasma*; Slovak Technical University in Bratislava: Bratislava, Slovakia, 2011.
39. Šimek, M.; Pekárek, S.; Prukner, V. Influence of power modulation on ozone production using an AC surface dielectric barrier discharge in oxygen. *Plasma Chem. Plasma Process.* **2010**, *30*, 607–617. [CrossRef]

MDPI
St. Alban-Anlage 66
4052 Basel
Switzerland
Tel. +41 61 683 77 34
Fax +41 61 302 89 18
www.mdpi.com

Coatings Editorial Office
E-mail: coatings@mdpi.com
www.mdpi.com/journal/coatings

www.ingramcontent.com/pod-product-compliance
Lightning Source LLC
Chambersburg PA
CBHW062035220926
51850CB00010B/2553
9783036509020